9th International Dry Dock Conference 2017

Houston, Texas, USA
5 - 6 May 2017

ISBN: 978-1-5108-4401-8

Printed from e-media with permission by:

Curran Associates, Inc.
57 Morehouse Lane
Red Hook, NY 12571

Some format issues inherent in the e-media version may also appear in this print version.

Printed by Curran Associates, Inc. (2017)

DM Consulting
c/o Joe Stiglich
12316 Dormouse Road
San Diego, CA 92129
USA

Phone: (858) 705-0760
Fax: (858) 538-5382

joe@drydocktraining.com

Additional copies of this publication are available from:

Curran Associates, Inc.
57 Morehouse Lane
Red Hook, NY 12571 USA
Phone: 845-758-0400
Fax: 845-758-2633
Email: curran@proceedings.com
Web: www.proceedings.com

TABLE OF CONTENTS

Financial Feasibility, Acquisition, Transportation & Operation of a Floating Dry dock

1. Introduction:

The general condition of Dry docks available for repair of commercial vessels in United States are in need of replacement. One of the options available for quick replacement of the dry dock capacity is to look for good condition dry docks overseas. As most of the dry docks are available in the Far East there are a lot of nuances to deal with in getting a dry dock.

United States is the biggest economy of the world and has the large number of vessels calling its shore but only a very small negligible fraction of the vessels carry out their repairs in United States. In fact even Jones Act vessels have been going overseas like the Bahamas to carry out their repair work. To a great extent this is due to lack of good dry docking and repair facilities in the United States. This has resulted in a loss of opportunities for US workers.

This paper is an attempt to highlight the various steps and precautions that needs to be taken in successfully bringing a floating dry dock from overseas to United States. To understand this a look into how a 12,000 TLC dock was successfully installed by a US company in the Gulf of Mexico is a good example as a case study.

2. History of VT Halter Marine

VT Halter Marine (VTHM) was bought out of bankruptcy in 2002 by VT Systems a Virginia based company which is a US subsidiary of a Singapore based company ST Engineering. VTHM is a reputed shipbuilder and has built new vessels and wanted to expand into providing a turnkey service that includes Ship Repair to its customers. VTHM started looking at various options for it to expand into Ship Repair Operations. The cost and time for a new built dock either in USA or overseas eliminated the option for a new dock. The alternative option was to get an existing floating dry dock as that will enable VTHM in leapfrogging to start providing the repair services in a shorter time.

3. Scouting for dock

There are a lot of brokers that will provide you a list of docks that are available in the market. Some of the well-known brokers are Howard Haffee located in Florida USA and Chris Murthy located in London UK. They generally have a pulse of the universe of the floating dry-docks available in the market and they are a good place to start.

You can narrow down your search based on lift capacity and size of the dock to suit your geographical location and customer base.

4. Identifying the dock

As the search is narrowed down to a specific dock.

The following drawings should guide in your decision making process.

General Arrangement & Lift Capacity (Fig 1)

Strength Calculations & Construction plan (Fig 2)

Electrical drawings.

Ballast Piping Arrangement

Bilge & Firefighting piping

Examples of the drawings for the 12,000 TLC dock are shown below

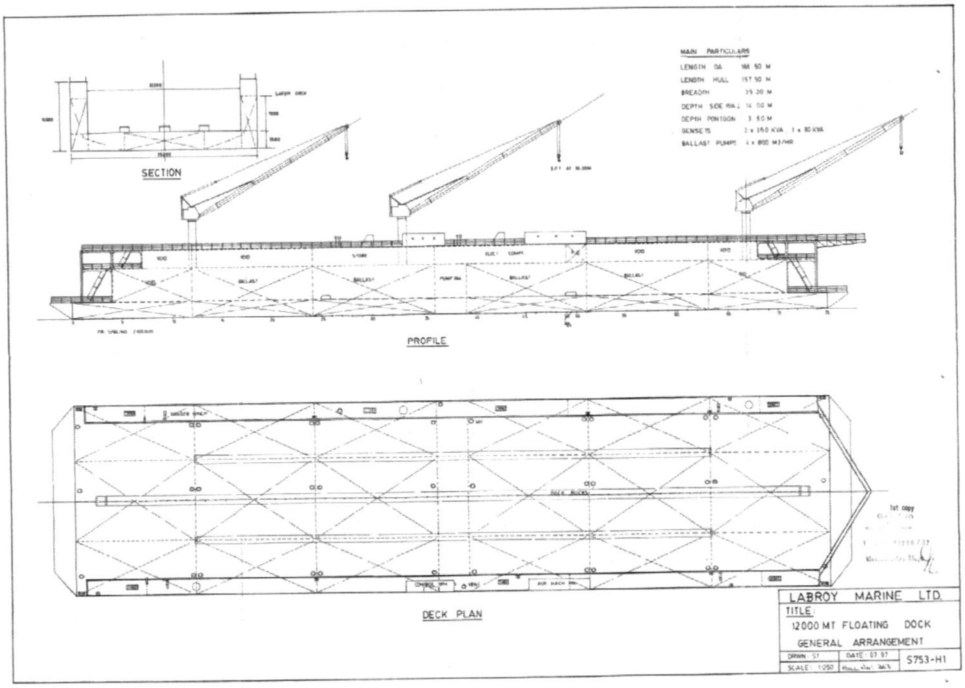

Fig 1: General Arrangement

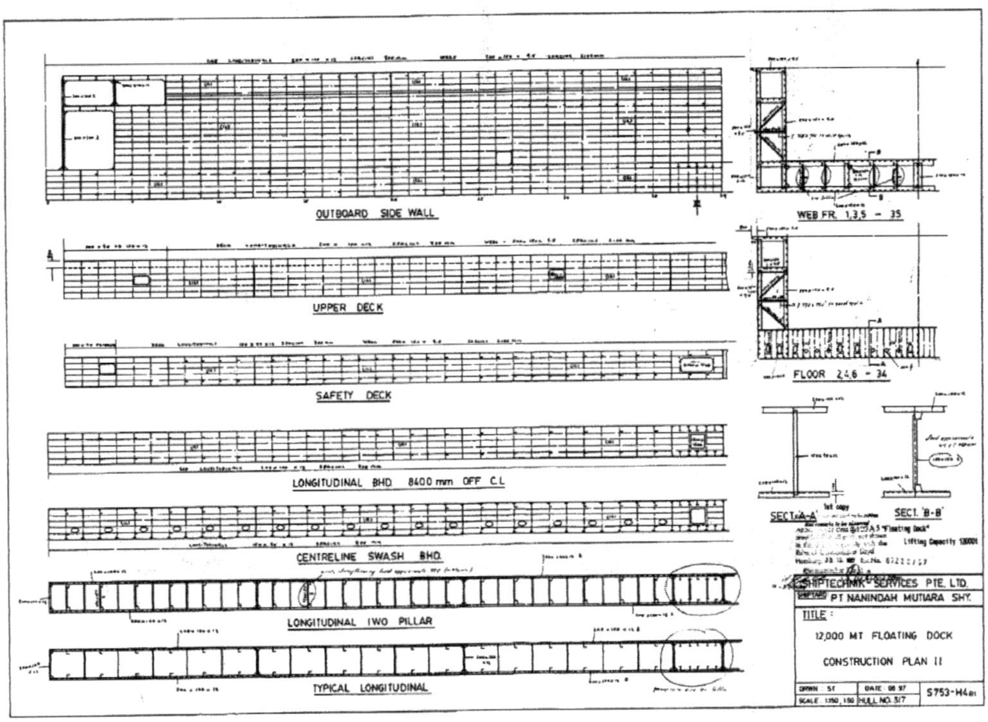

Figure 2: Construction plan

5. Class certification

If the dock is certified and the certification is current it gives a good gauge on the condition of the dock as there has been an independent third party verification. The certification can be either by any classification society or by a private third party inspection service provider. You will find that the docks that are put up for sale either have their certification expired or not much has been done to maintain the upkeep of the dock.

Example of a dock certification is shown in Figure 3

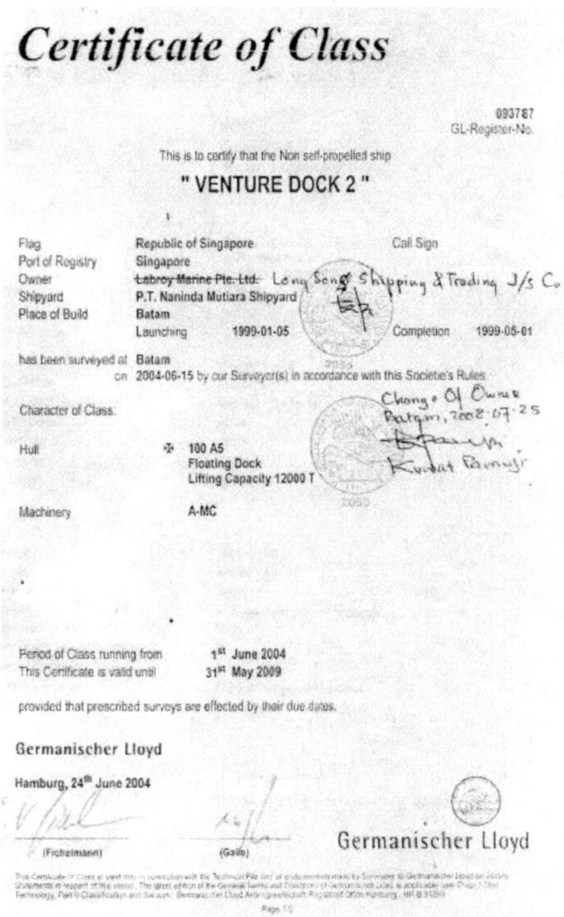

Certificate of Class

093787
GL-Register-No.

This is to certify that the Non self-propelled ship

" VENTURE DOCK 2 "

Flag: Republic of Singapore   Call Sign
Port of Registry: Singapore
Owner: ~~Labroy Marine Pte. Ltd.~~
Shipyard: P.T. Naninda Mutiara Shipyard
Place of Build: Batam
Launching 1999-01-05   Completion 1999-05-01

has been surveyed at Batam
on 2004-06-15 by our Surveyor(s) in accordance with this Societie's Rules

Character of Class:

Hull: ✠ 100 A5
Floating Dock
Lifting Capacity 12000 T

Machinery: A-MC

Period of Class running from 1st June 2004
This Certificate is valid until 31st May 2009

provided that prescribed surveys are effected by their due dates.

Germanischer Lloyd

Hamburg, 24th June 2004

(Fichelmann)   (Galle)

Germanischer Lloyd

Figure 3: Dock Certification

6. Offer Letter

In this stage of the process you have already located the dock that will suit your requirement and have looked at the drawings and have an idea as to what you would be willing to pay for the dock. It is recommended that the dock be physically inspected to get a better gauge on the condition to the dock but you can make your offer subjective to satisfactory due diligence.

The offer letter at a minimum should address the following points:-

A) Purchase Price.

Price to pay on a as is basis subject to satisfactory due diligence. Cost for transportation and any other taxes should be taken into account before deciding on the purchase price, this is called the landed cost of the dock.

Landed Cost = Purchase price + Transportation + Taxes (if any)

B) Deposit.

The offer letter need to specify if a deposit is to be made and time required to make a deposit. Generally it is about 7-14 business days after signed acceptance of the offer and the amount is about 10% of the purchase price.

C) Inspection:

It is good practice to inspect the dock before making an offer. Generally inspection can be done by a third party who has the expertise and there are a lot of inspection agencies and one of them is Heger' Dry dock Inc from Holliston, MA. Special attention is to be paid to the thickness of the bottom plate and the condition of the paint of the underwater hull, because due to the size of the floating dock it is difficult to find a place to dock the floating dock especially in USA.

D) Delivery:

The place of delivery need to be specified in the offer letter. Generally the docks are offered on a as is basis.

E) Condition of Delivery:

The condition of delivery need to specify that the sale is free of all charters, encumbrances, mortgages, liens or any other debt whatsoever.

F) Condition for towing/transportation:

It is good practice to specify that the dock need to be prepared to be ready for towing, if wet tow is planned or ready for loading on to a heavy lift vessel/barge, if a dry tow is planned. The towing preparation to be prepared to the satisfaction of the buyers towing surveyor. Having this clause eliminates the need to hire local people at the country the dock is in. Being a foreigner we may not have the resources to execute the tow preparation nor be able to estimate the cost involved. There may be rules or condition specific to that country which may also not be aware.

G) Approval by Board Of Directors (BOD) or Authorities

If the company requires BOD approval and that has not been obtained that needs to be specified in the letter that the offer is subject to buyers BOD. The seller needs to obtain all permissions and approvals required in the seller's country.

7. Memorandum of Agreement

Once the sellers and buyers have agreed and accepted the offer price and conditions. A Memorandum of agreement needs to be signed. In our transaction we had used the BIMCO SALE FORM 1993. (Figure 4)

Norwegian Shipbrokers' Association's Memorandum of Agreement for sale and purchase of ships. Adopted by the Baltic and International Maritime Council (BIMCO) in 1956.
Code-name
SALEFORM 1993
Revised 1966, 1983 and 1986/87.

MEMORANDUM OF AGREEMENT

Dated: *May ___,2012*

Long Song Shipping and Trading JSC of, 135/17/23, Nguyen Huu Canh Street, Ward 22, Binh Thanh District, Ho Chin Minh City, Vietnam

Figure 4: BIMCO SALE FORM 1993.

8. BILL OF SALE

An official Bill of Sale is a document that is required to show proof of transfer of title. Example of the Bill of Sale is shown in Figure 5.

Prescribed
By the commissioners
Of Customs & Excise
With the consent of the
Secretary of State for
Trade and Industry

Form No.11 A

X.S. 79 A

BILL OF SALE

Official Number	Name	Year Built and port of registry	Type	Propelling power of engines (KW)
24800899	Venture Dock II	1999, Ulaanbaatar	Floating Dock	Nil

	Meters	Number of Tons (where dual tonnages are assigned the higher of these should be stated)	
		Gross	Net
Length Over All	166.50		
Registered Breadth	39.2	9,255 Tons	2,777 Tons
Registered Depth	14.00		

And as described in more detail in the Register Book

We, (a) [redacted] (hereinafter called "transferor's") having our principle place of business at No.135/17/23, Nguyen Huu Canh Street, Ward 22, Binh Thanh District, Ho Chi Minh City, Vietnam in consideration of the sum of USD[redacted]y) paid to us by (b) [redacted] (hereinafter called "the transferee(s)") of 900 Bayou Casotte Parkway, Pascagoula, MS 39581 USA, the receipt whereof is hereby acknowledged, transfer all the shares in the Ship above particular described, and in her boats and appurtenances, to the said transferee(s).

Transferors hereby confirm that they are the holders of one hundred percent (100%) of the shares in the Ship which shares shall be transferred to the transferees upon receipt of the full purchase price.

Further, we, the said transferors for ourselves and our successors covenant with the said transferee(s) and (c) **their assigns**, that we have power to transfer in manner aforesaid the premises hereinbefore expressed to be transferred, and that the same are free from all encumbrances (d) **mortgages, maritime liens, claims, taxes or any other debts whatsoever**.

In witness whereof we have duly executed this Bill of Sale on ______ Sept 2012.

For Long Son Shipping & Trading Joint Stock Company

Name: Tran Minh Quang
Title: President

(a) Name in full of Body Corporate (b) Full name (s) and address(es) of transferee(s) with their description in the case of individuals, and adding "as joint owners" where such is the case (c) "his", "their" or "its". (d) I any subsisting incur add "save as appears by the registry of the said ship" (e) Signatures and description of witnesses, i.e. Director, Secretary, etc. (as the case may be).

Figure 5: Bill of Sale

9. Provisional Certification

Before the transfer of title is done a provisional certificate of class need to be acquired and the requisite forms should be filled well in advance before the actual transfer of title is done. Full-fledged classification society have a more rigorous requirement both for inspection and documentation. As the purpose of getting classification was for the transition period then use of alternative classification society like Mongolia Maritime Administration was used and the documentation and inspection was faster and less stringent.

The following documents are required.

a) Bill of Sale (Figure 5)

b) Declaration of ownership (Figure 6)

c) Application to Register (Figure 7)

MONGOLIA

MONGOLIA MARITIME ADMINISTRATION

Declaration of Ownership
(Form M2)

Chapter 3 of the Regulations for Registration of ship 2003

Mongolia Ship Registry Pte Ltd
133 New Bridge Road
#16-02 Chinatown Point
Singapore 059413
Tel: (65) 6225 0125
Fax: (65) 6225 0305
Email: operations@mngssl.org
Website: www.mngssl.org

A. SHIP'S PARTICULARS

Ship Name (for registration) VT HALTER MARINE DOCK 1	Type of Ship Floating Dock	Year Built 1999	Keel Laid -
Place / Country Built Batam, Indonesia	IMO Number -	Gross Tonnage 9255	National Gross Tonnage
Net Tonnage 2777	National Net Tonnage	Deadweight	Hull Material (steel, etc) Steel
Length of Ship (Metres) LOA: 166.50 / ITC-69: 151.20	Breadth (Metres) 39.20	~~Moulded~~ Depth (Metres) 14.00	Moulded Draught (Metres)
Builder's Name P.T. Naninda Mutiara Shipyard			Previous Registry Ulaanbaatar
Previous Owner Long Son Shipping & Trading J/S Co.			Ship Name (In Previous Registry) Venture Dock 2

B. OWNER'S PARTICULARS

Full name(s)	Full address(es)	~~Number of shares~~ Interest owned
VT Halter Marine, Inc.	900 Bayou Casotte Parkway Pascagoula, MS 39581	100%
	Total number of shares in ship	100 %

Nature of interest*: ☒ Sole ownership ☐ Joint ownership ☐ Ownership in severalty

C. DECLARANT'S PARTICULARS

Full name(s)	Address	Nationality
[redacted]	900 Bayou Casotte Parkway Pascagoula, MS 39581	USA

Status of declarant*:

☐ Director of owning corporation ☐ Secretary of owning corporation

☐ Individual/joint owner(s) ☒ Authorised person

I/We*, whose name(s) is/are* hereunto subscribed, hereby declare that:

.1 I am/we are* duly authorised to make this declaration;

.2 all the particulars stated hereon and in the application form dated .... are correct;

.3 ~~the property in the ship is divided into 100 shares~~; and

.4 no person, other than those mentioned in section B is/are* entitled to be registered as owner(s) of the ship,

And I/we* make this solemn Declaration conscientiously believing the same to be true and by virtue of the Statutory Declaration Act (Cap 320).

Name(s) and signature(s) of declarant(s)	Declared before me on (date):

Page 1 of 2

Figure 6- Declaration of Ownership

MONGOLIA
MONGOLIA MARITIME ADMINISTRATION

Application to Register a Ship
(Form M1)

Chapter 3 of the Regulations for Registration of ship 2003

Mongolia Ship Registry Pte Ltd
133 New Bridge Road
#16-02 Chinatown Point
Singapore 059413
Tel: (65) 6225 0125
Fax: (65) 6225 0305
Email: operations@mngssl.org
Website: www.mngssl.org

A. SHIP'S PARTICULARS

Ship Name (for registration) **VT HALTER MARINE DOCK 1**	Type of Ship **Floating Dock**	Year Built **1999**	Keel Laid -
Place / Country Built **Batam, Indonesia**	IMO Number of Ship -	Gross Tonnage **9255**	National Gross Tonnage
Net Tonnage **2777**	National Net Tonnage	Deadweight	Hull Material (steel, etc) **Steel**
Length of Ship (Metres) ⓟ LOA **166.50** / ITC-69 **151.20**	Breadth (Metres) **39.20**	~~Moulded~~ Depth (Metres) 14.00	Moulded Draught (Metres)
Builder's Name **P.T. Naninda Mutiara Shipyard**			Previous Registry **Ulaanbaatar**
Previous Owner **Long Son Shipping & Trading J/S Co.**			Ship Name (in Previous Registry) **Venture Dock 2**

B. SHIP VOYAGE INFORMATION

Sea Area	Trading Area
For special registration, please fill in the following information.	
Voyage Limitations	
Port of Departure	Port of Arrival
Estimate Date of Departure	Estimate Date of Arrival

C. OWNER'S PARTICULARS

Name **VT Halter Marine, Inc.**	IMO Number of Owner -
	Name of Person in Charge **William E. Skinner**
Address **900 Bayou Casotte Parkway Pascagoula, MS 39581**	Telephone **+1 228-696-6865**
	Facsimile **+1 228-696-6890**
	E-mail **b.skinner@vthm.com**

ⓟ *Length of ship (LOA) – Overall length of the ship.*
Length of ship (ITC-69) – Length of the ship as defined under International Tonnage Convention

Page 1 of 6

Figure 7: Application to register

After the documentation required is provided and the requisite fees is paid then a provisional certificate is issued. (Figure 8)

MONGOLIA

MONGOLIA MARITIME ADMINISTRATION

PROVISIONAL CERTIFICATE OF REGISTRY

Issued under the provisions of
Chapter 3 of the Regulations for Registration of Ships 2003

Official Number	Call Sign	IMO Ship Number
24B00899	-	-

Name of Vessel
VT HALTER MARINE DOCK 1

Name and Address of Owner	IMO Registered Owner Number	Shares (%)
VT HALTER MARINE INC. 900 Bayou Casotte Parkway, Pascagoula MS 39581. USA	6090179	100

Description of Vessel

Type of Vessel	Year of Built	Hull material	Previous Registry / Name	
Floating Dock (NP)	1999	Steel	- / VENTURE DOCK 2	
Gross Tonnage	**Length(Metres)**	**Name of Builder**		
9255	LOA - 168.50 ITC - 151.20	PT. Nanindra Mutiara Shipyard, Indonesia		
Net Tonnage	**Breadth(Metres)**	**Number of Engines**	**Type of Engine / Total Power (KW)**	
2777	39.20	-	- / -	
Deadweight	**Depth(Metres)**	**Speed of Vessel (Knots)**	**Engine Make / Model.**	
-	14.00	-	-	

Date of Provisional Registry	Port of Registry	Date of Expiry
22 October 2012	Ulaanbaatar	21 April 2013

Issued at Singapore on 22 October 2012

The Registrar of the Mongolia Ship Registry, by the powers vested thereupon by the Regulations for Registration of Ships 2003, hereby authorizes this Provisional Registration of Vessel **VT HALTER MARINE DOCK 1**

MONGOLIA
SHIP REGISTRY

BAZARRAGCHAA ALTAN-OD
REGISTRAR
MONGOLIA SHIP REGISTRY
MONGOLIA MARITIME ADMINISTRATION (OPERATIONS)

MSR/PROV CR/REV 3/2010

Figure 8- Provision Certificate of Registry

10. Towing of the floating dock (Wet Tow)

A towing plan is prepared for the wet tow of the dock from Vietnam to Philippines for carrying out repairs of the dock. It may or may not be necessary for this intermediate tow depending on the condition of the floating dock. A typical wet tow plan is shown in Figure 9.

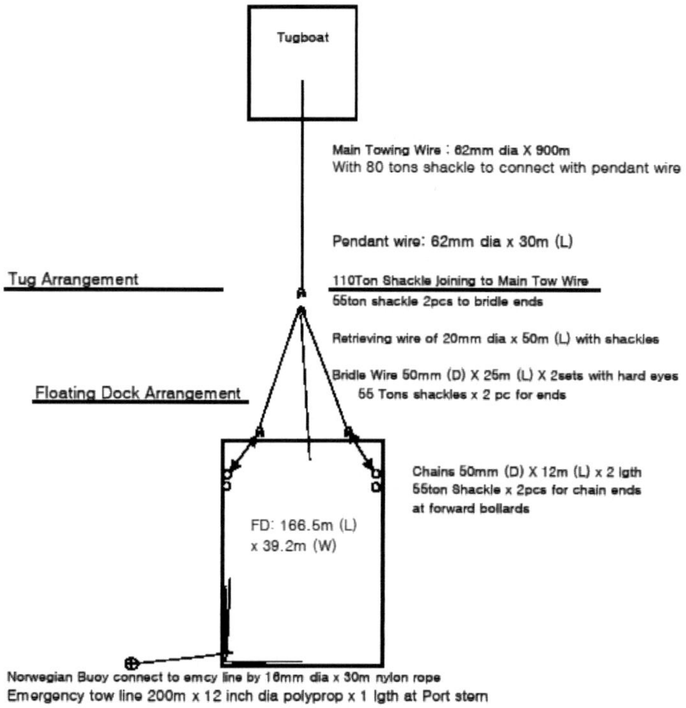

Towing gears required (approx):
1) Chaffing chains: 50mm dia x 12m x 2 lgths & 4 kenter shackles
2) Bridle wires: 50mm dia x 25m (L) x 2 pcs with both ends hard eyes for 55 tons SWL shackles
3) Retrieving wire: 20mm dia x 50m (L) with 2 x 20 tons shackles
4) Emergency towing line: 200m x 12 inch circumference polyprop line with end spliced eyes
5) Buoy with 16mm dia x 30m (L) nylon rope to polyp line, end eyes spliced
6) Shackles: At chains/Bridle wires – 55 tons SWL x 10 pc
At emergency towing line – 80 tons x 1 pc
7) 25mm dia wires x approx 200m (L), shackles, clips, etc to secure anchor chains, dock cranes,
8) Ropes approx 30mm dia x 100m for securing loose items as required
9) Shackles of 20 tons & 35 tons as required
10) Navigation lights to be fitted at Forward P/S & stern

Figure 9- Typical Wet Tow plan.

In this case it is a good learning experience to show what goes into the preparation for a wet tow. A survey is required to determine the suitability of the tow.

Generally it is a twostep process the surveyor comes to inspect the dock and advises on the preparation that is required to prepare for towing. He looks at the towing point and towing arrangement. He also looks at the lashing required for the loose gear and will or will not require a tow longitudinal and transverse strength calculation. He provides a list of recommendation that need to be done to meet the tow and provides the tow plan. Before the vessel is towed he will come and inspect and certify that the preparation done is satisfactory. This will be required both by the towing company and also for insurance purposes. A copy of the stability report is attached in Appendix A.

Some of the preparations done for the wet shown are shown in Figures 10A to Figure 10D

1) Forward port & Stbd side double bollard each 400 mm dia to be renewed including the foundation base to main deck, plus a pad eye.

Corroded double bollards and pad eye for towing point to renew

Figure 10A- Wet Tow preparations.

Figure 10B- Wet Tow Preparations

Figure 10C- Wet tow preparations

To renew one set of access ladder from main deck to side walls, safety deck

Figure 10 D- Wet Tow preparations

11.Dock Refurbishing:

If the floating dock is being bought from the Far East it is recommended that the floating dock be dry-docked and inspected prior to it being towed to USA. There are more options of docking in the Far East and also the pricing is better. Certainly this will depend on the condition and age of the dock. A new coat of paint and new set of anodes will always pay off in the long run. Another aspect that is critical is the sea valves. It is good common sense to get the sea valves opened up cleaned and inspected while the vessel is being dry-docked. Advantage of the dry-docking can be done to take care of other aspects as required based on individual case to case and condition of the dock. Figures 11A to Figure 11D show the various stages of the dock being refurbished.

Figure 11A- Dock Refurbishing- Hull sides blasted to pure white metal.

Figure 11B- Dock Refurbishing- Dock Deck wing walls blasted to white metal.

Figure 11C- Dock Refurbishing- Under deck painted.

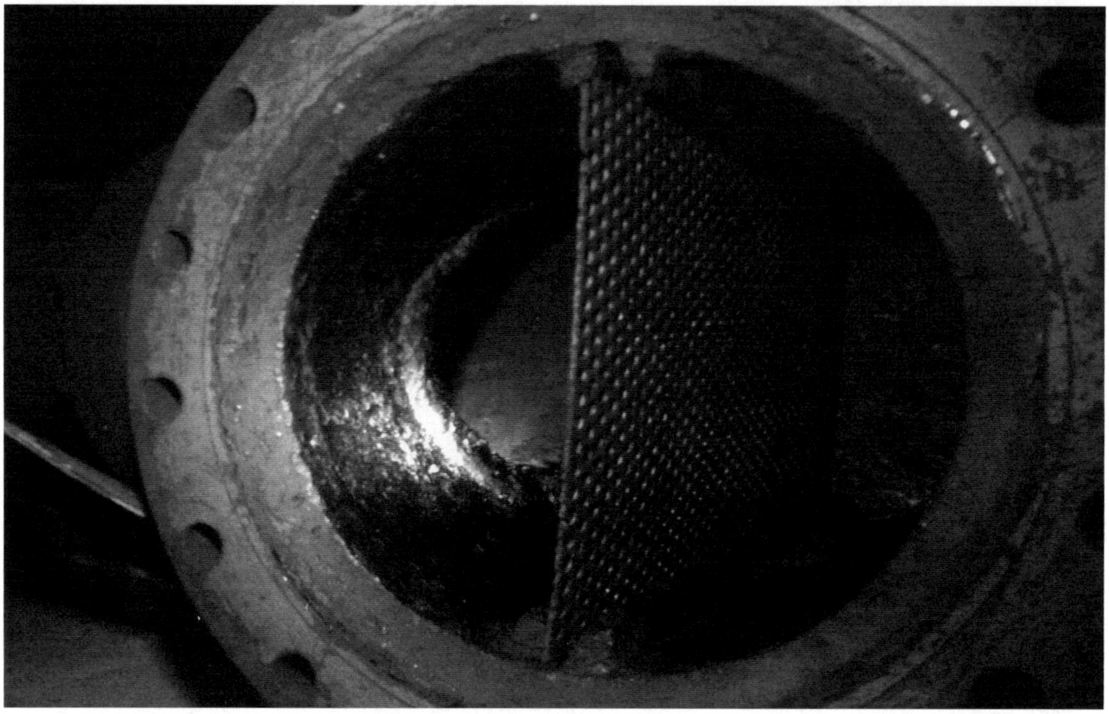

Figure 11D- Dock Refurbishing- Sea Strainers cleaned and coated.

A feature that was installed on the dock was to create a stub piece for each opening and a temporary blank prepared for each opening. This was an emergency feature created to be able to install the blank by divers and this will provide access to carry out any pipe or valve repair if required in the future as we know that finding another dock is both difficult due to availability of the docking capability and also time and money. Figure 12A – Figure 12B show the stub pipe for the sea valves.

Figure 12A- Sea Valves stub pieces side view

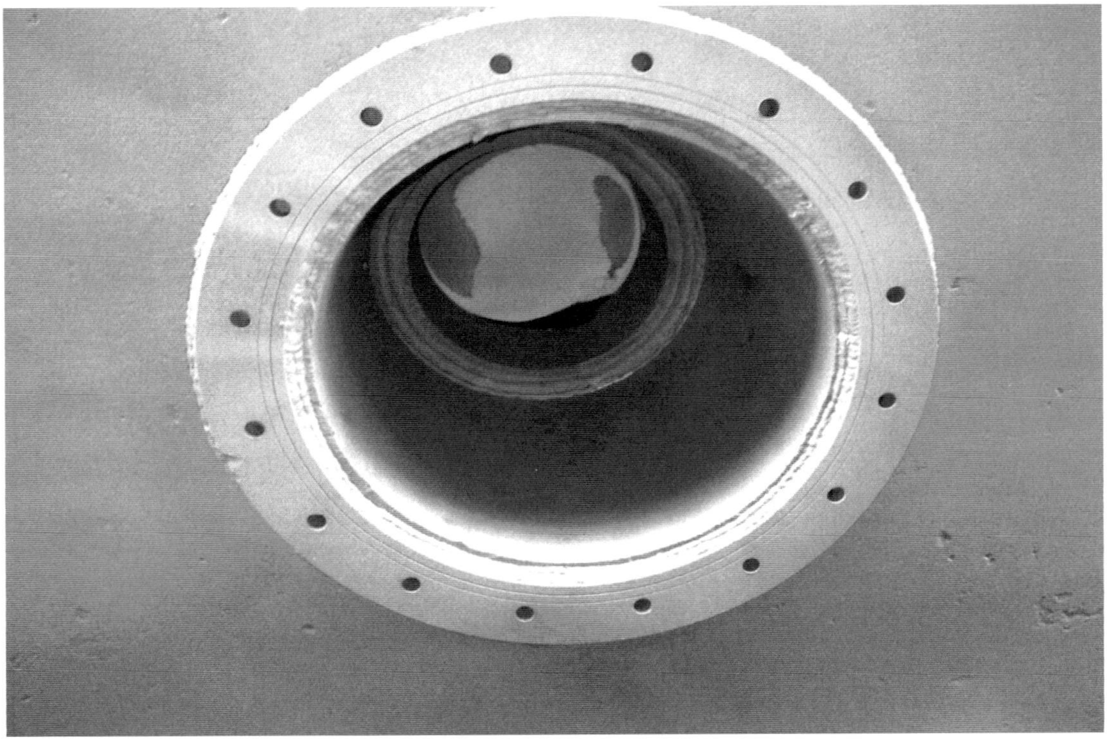

Figure 12B- Sea Valve stub pieces front view

12. Document Check List:

The following documents are required before the dock can be bought into USA territory.

A) Registry Certificate

B) Letter of Attestation- Exemption from Load Line

C) Interim International Tonnage Certificate

D) International Antifouling certificate

E) Safety Construction Certificate

These certificates are issued by the class surveyor once the refurbishment is completed. In this case the certificates was issued by Sing class Singapore which was authorized by Mongolia Maritime Administration.

13.Transportation of the Floating dock to USA – Dry Tow

The dry tow portion needs to be planned well in advance as the cost of the transportation will depend availability of the heavy lift vessel and if it can be planned such that the heavy lift vessel does not need to do a lot of deviation and if any leg of the round trip is not empty under ballast. There are a limited number of heavy lift vessel and depending on market conditions the rate can vary. Also the cost of the fuel will be a factor and is generally baselined both in terms of quantity and the cost at the time of operation.

BIMCO HEAVY CON 2007 was used as a contract for the Dry tow. (Figure 13) A full copy of the BIMCO HEAVY CON is attached in Appendix B.

BIMCO

HEAVYCON 2007

STANDARD HEAVYLIFT CHARTER PARTY

PART I

Figure 13- BIMCO HEAVY CON 2007

A Transport Manual should be requested and is generally provided for the heavy lift company and has the following key items:

1) Schedule of Loading & Offloading (Fig 14A & 14B)
2) Load On planning & Load off Planning (Fig 15A & 15B)
3) Stowage of the Cargo (Floating dock) on the vessel and sea fastening. (Fig 16A & Figure 16B)
4) Sea Route.(figure 17)

A full transport Manual is attached in Appendix C.

Schedule—Dry Transportation of 1 Dock

DATE	CONTENTS	4	5	6	7	8	9	10	11	12	13	14	15	16	17	18	19	20	21	22	23
	LOADING																				
Day 1	ZHENHUA 29will arrive at specified anchorage of Philippines, anchor and pre-ballast.					■	■	■	■												
Day 2	1. Ballasting the vessel to the -16.3m.		■	■	■	■	■	■													
	2. Preparations for the loading.						■	■													
	3. Loading the Dock.								■												
	4. Unmooring the lines of tugs after the 1Dock's position.									■											
	5. Deballasting the vessel to the summer load draft.										■	■	■	■	■	■	■	■	■		
Day 3-5	Sea-fastening.																				
Day 6	ZHENHUA 29 sails.																				

(work time(an hour/grid))

Figure 14 A- Schedule for Loading On heavy lift vessel.

	OFF-LOADING																				
1st day	1. ZHENHUA 29 will arrive at designated anchorage of America, anchor and pre-ballast.																				
	2. All the sea-fastenings between the vessel's deck and the Dock will be cut loose and removed.									■	■	■	■	■	■						
2nd day	1.Ballasting the vessel to -16.3	■	■	■	■	■	■														
	2. Preparation for the off-loading(tugs mooring etc.)						■	■													
	3.Off-loading the Dock								■												
	5. The DOCK will be towed to the specified location.										■	■	■								
	6. Deballasting the vessel to the summer load draft.										■	■	■	■	■	■	■	■	■	■	■
3rd day	ZHENHUA 29 sails.																				

Note:The schedule will be executed under the normal weather condition. If there are some unfavorable factors, such as wind speed exceeding 8m/s /s or wave height exceeding 0.5m, the schedule may be delayed.

Figure 14B- Schedule for Off Loading from Heavy Lift vessel

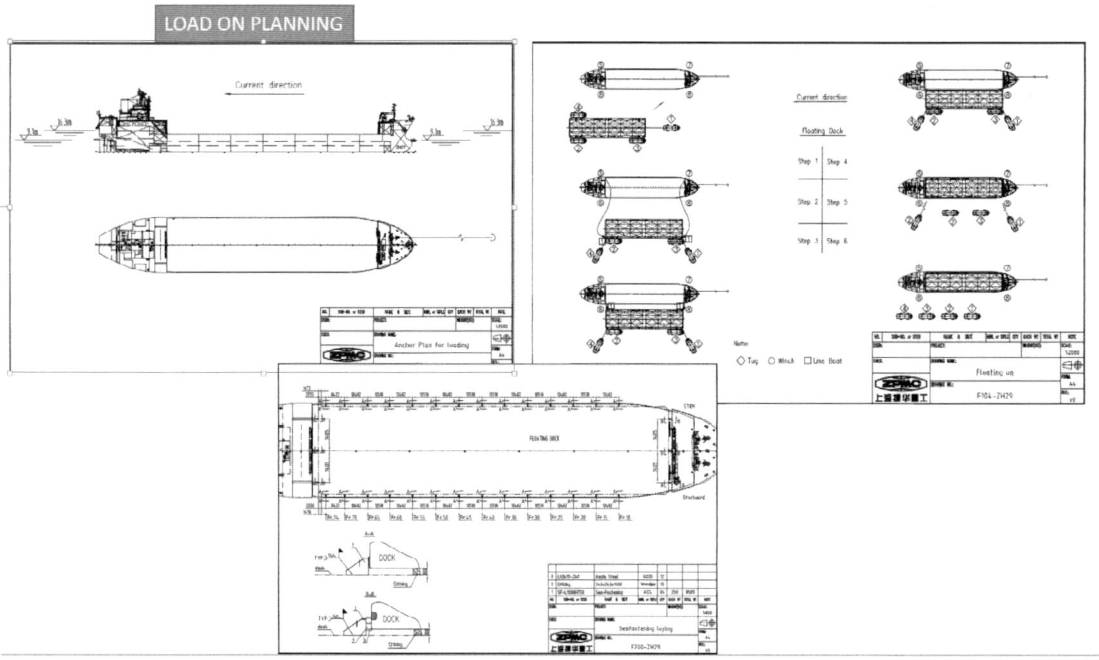

Figure 15A- Showing the Load On sequence

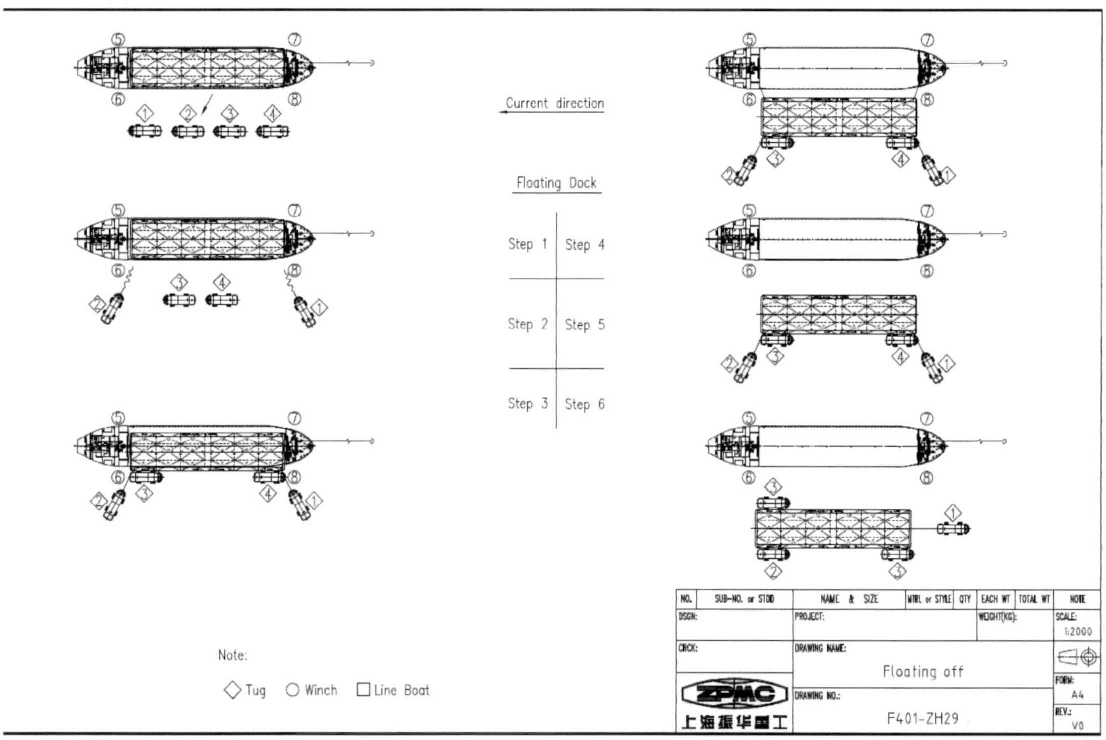

Figure 15B- showing Load off sequence

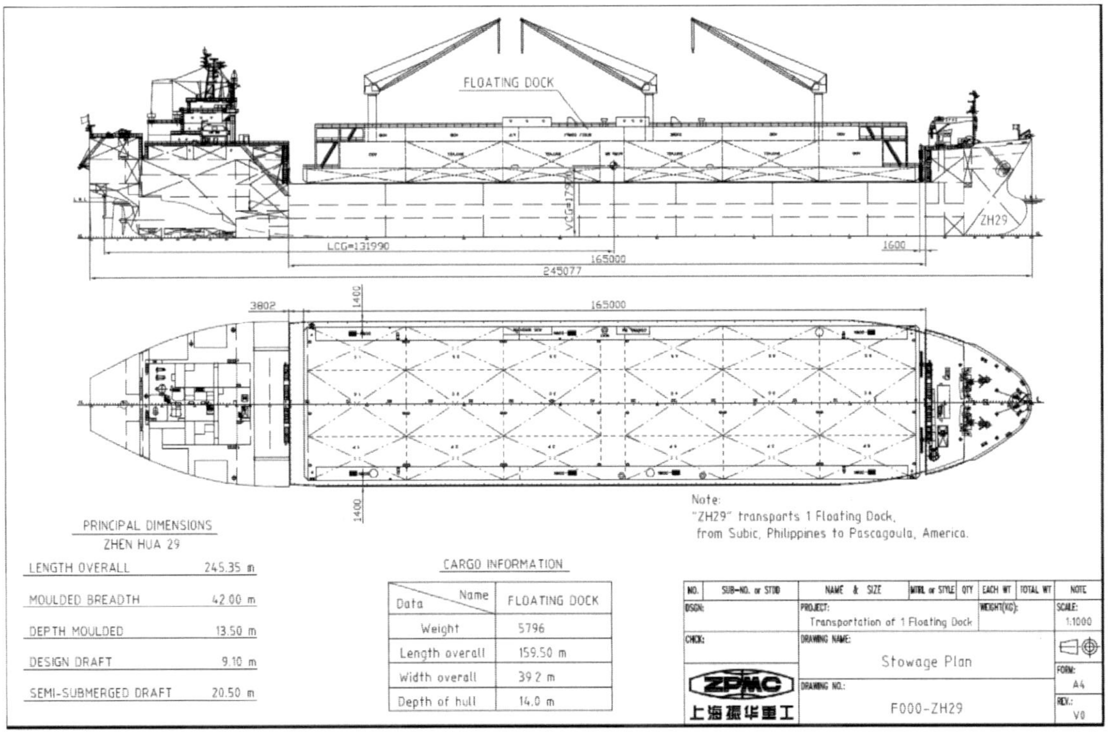

Figure 16A- Showing stowage of the Cargo (Floating Dock) on the heavy lift vessel

Figure 16B- Picture of floating dock on the heavy lift vessel

5.3 Sea Route

Vessel	ZHENHUA 29	Port of departure	Subic, Philippines
Condition	Voyage Departure	Via	/
Data source	Global Wave Statistics Online	Port of destination	Pascagoula, America

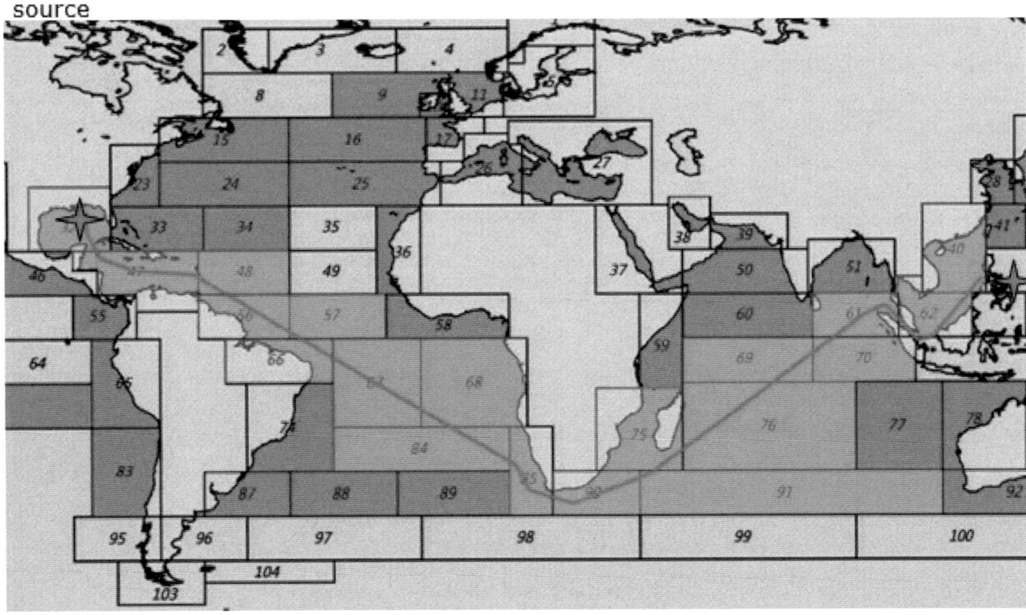

WIN Company (Japan) will offer the information of weather routing all along the voyage. ZHENHUA 29 should evade the unfavorable sea area that the parameters of weather routing exceed to set-values.

Figure 17- Planned Sea Route

14.Financial Feasibility Analysis:

The financial feasibility analysis is discussed in this section. This is generally done before you even procced to acquire the floating dock to ensure that the project is financially feasible. The financial feasibility has the following key components.

1. Capital Outlay: This is the entire cost for the project. This includes the cost for the dock, including transportation and refurbishment and taxes and fees if any. If there are any civil improvements that need to be done like dredging, bulkhead, shops, offices and equipment.
   For the example shown it was $45M.
2. Revenue stream period: This is the period for which the financial feasibility is done. It can vary from 15 years to 20 years. In this case the period chosen was 20 years.
3. Discount Rate: This is the cost of money. This is based on the assumption that the capital even if invested from existing funds and not borrowed fund there is an associated cost of money. Depending on the standing of the owners and type of owners this can vary. In this case 5% was used as the discount rate or cost of money.
4. Margin: This is the expected operating profit that the repair projects will provide. It is the Gross Margin before Sales General & Administration (SG&A).
5. Terminal Value: This is the value of the assets at the end of the analysis period. It can be either calculated by using the scrap value of the assets or it can be assed as a going concern and use multiples of the cash flow. In this case a $10M scrap value was assumed for the dock.
6. Discounted Factor: This calculates the present value of the future year cash flow and this is related to the discount rate used. If the discount rate is higher the present value will be lower and if the discount rate is lower the present value will be higher.
7. Discounted Annual Cash Flow: This the present value of the cash flow based on the estimated annual cash flow of the business in future years.
8. Cumulated Discounted Cash flow: This is the sum of all cash flows. It is basically negative in the initial years as you pay down the initial investment and then positive in the future years. In this example it shows that in Year 7 the amount is positive.

9. Internal Rate of Return (IRR): This is the % of the return on your capital invested. If it is based on the Annual Cash flow it is called IRR-Annual Cash Flow. If it is based on discounted cash flow it is called IRR- Discounted Cash Flow.

A copy of the financial analysis is shown in Figure 18. An excel copy is attached in Appendix D for easy reference to adjust the numbers for future use.

Financial Feasibility Analysis

1	**Capital Outlay**		
	Improvements	**45**	
	Total	**45**	
2	Revenue per annum for	20	years
3	Discount Rate	5.00%	
	Margin	25%	

		Yr 0	Yr 1	Yr 2	Yr 3	Yr 4	Yr 5	Yr 6	Yr 7	Yr 8	Yr 9	Yr 10
	Initial Outlay	(45)	-	-	-	-	-	-	-	-	-	-
	Estimate Revenue		20	30	35	40	40	40	40	40	40	40
	Margin		5	8	9	10	10	10	10	10	10	10
	Terminal Value											
	Annual Cashflow	**(45)**	**5**	**8**	**9**	**10**	**10**	**10**	**10**	**10**	**10**	**10**
	Discount Factor	1.000	0.950	0.903	0.857	0.815	0.774	0.735	0.698	0.663	0.630	0.599
4	Discounted Annual Cashflow	(45)	4.75	6.77	7.50	8.15	7.74	7.35	6.98	6.63	6.30	5.99
5	Cumulative Discounted Cashflow (CDC)	(45.00)	(40.25)	(33.48)	(25.98)	(17.83)	(10.10)	(2.75)	4.24	10.87	17.17	23.16

		Yr 11	Yr 12	Yr 13	Yr 14	Yr 15	Yr 16	Yr 17	Yr 18	Yr 19	Yr 20
	Initial Outlay										
	Estimate Revenue	45	45	45	45	45	50	50	50	50	50
	Margin	11	11	11	11	11	13	13	13	13	13
	Terminal Value										10
	Annual Cashflow	**11**	**11**	**11**	**11**	**11**	**13**	**13**	**13**	**13**	**23**
	Discount Factor	0.569	0.540	0.513	0.488	0.463	0.440	0.418	0.397	0.377	0.358
4	Discounted Annual Cashflow	6.40	6.08	5.78	5.49	5.21	5.50	5.23	4.97	4.72	8.07
5	Cumulative Discounted Cashflow (CDC)	29.56	35.64	41.42	46.90	52.11	57.62	62.84	67.81	72.52	80.59

6	IRR based on Annual cashflow	19.3%
	IRR Based on Discounted Cash Flow	9.7%

Figure 18-Financial Analysis

Carrier Undocking Presentation

Improved Process For Undocking Aircraft Carriers After Extended Stays In Dry-dock

March 1, 2017

Adam J. Bierbauer – NNS Naval Architect
John R. Anderson – NNS Dockmaster

Carrier Undocking Presentation

Why Change?

- Reduce Cost - Old process was too labor intensive
- Improve Results - Old process was imprecise
- Minimize Risk - Maintain continuous control of ship's list

Significant weight changes

<u>Old Process was too Labor Intensive</u>:

- The process involved continuous tracking of repair items, temporary equipment and systems taken aboard, removed from, and relocated within the ship.
- Permanent alterations to the ship were accounted for by estimating percent completion of removals and installations identified by drawings.
- Required expenditure of thousands of man-hours by engineering, trades foremen, and ship's force over the duration of the dry-docking period.

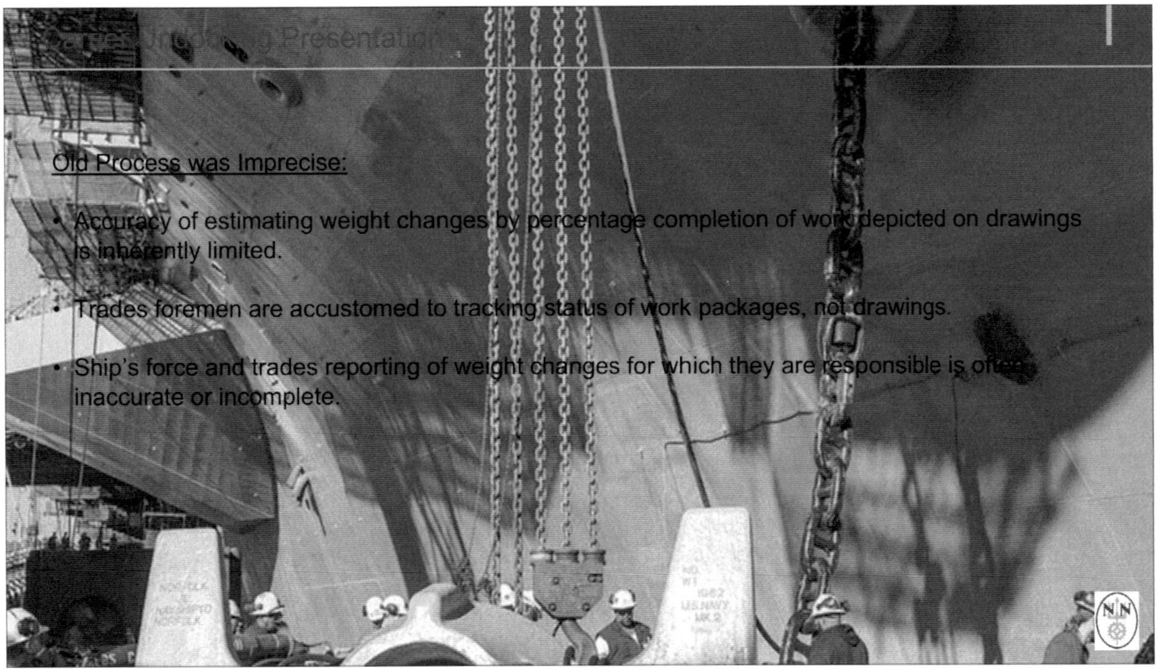

Old Process was Imprecise:

- Accuracy of estimating weight changes by percentage completion of work depicted on drawings is inherently limited.
- Trades foremen are accustomed to tracking status of work packages, not drawings.
- Ship's force and trades reporting of weight changes for which they are responsible is often inaccurate or incomplete.

Carrier Undocking Presentation

Minimizing Risk by Controlling Ship's List:

- During previous undockings, dry-dock flooding was not halted to make list corrections, resulting in ship floating before list correction was complete.
- During a carrier undocking in 1999, slow response in operating the list control system resulted in ship floating before list correction was even started.
- The improved process maintains continuous control of ship's list while the dock is being flooded.

Carrier Undocking Presentation

Undocking Aircraft Carriers At Puget Sound Naval Shipyard

•During modernization of a class of carriers in the 1950's, San Francisco Naval Shipyard employed a method developed at Puget Sound using gunners quadrants to detect impending heel and correct for it before the ships left the blocks.

•The quadrant method is still used at Puget Sound, but the extreme precision achieved during the 1950's modernizations is considered unnecessary.

•Undocking ballast requirements and draft predictions are based solely on surveys of tank liquids, hangar bays, flight deck, and storerooms conducted upon entering and prior to leaving the dock.

•List when afloat of 1 degree or less is considered satisfactory.

•When the quadrant reading is ample enough to predict the direction of heel, flooding of the dock is stopped and sufficient weight is moved to result in a 1 degree list change in the opposite direction.

Carrier Undocking Presentation

Newport News Improved Process

- Eliminate continuous weight tracking effort while ship is in dock.
- Establish basic undocking ballast requirements based on historical data.
- Conduct surveys of tank liquids, hangar bays, flight deck and storerooms for comparison with previous overhauls and adjust basic undocking ballast if needed.
- Once draft on the hull reaches approximately 28 feet, dry-dock flooding proceeds at a controlled, slow rate.
- Use precision micrometer levels on deck and dockside transits to monitor list during flooding of dry-dock.
- List correction is accomplished using one or more transporters loaded with crane test weights on flight deck and, if necessary, ship's list control system.
- If list approaches 150 seconds and the ship is still on the blocks, stop dry-dock flooding and correct list by transferring weight from heavy side to light side of ship. Wait approximately 30 minutes to observe bubble reaction on micrometer level.
- After first list correction, resume flooding dry-dock at a slow rate. If list again approaches 150 seconds and the ship is still on the blocks, repeat the previous step.

Carrier Undocking Presentation

Historical Ballast Requirements

- Examination of previous complex overhauls (COH or RCOH) on several carriers has revealed that the ballast, consisting of liquids in tanks and hard ballast on flight deck, needed to produce the desired trim and list would have required the following moments:

	Required Moment Fwd of Amidship	Required Moment Port of Centerline
Min	807,000 ft-tons	48,000 ft-tons
Max	1,026,000 ft-tons	60,000 ft-tons

- Selecting values mid way between the Min and Max values results in possible trim and list variances, relative to the Min and Max values, of:

Trim	+/- 0.7 feet
List	+/- 0.4 degree

Carrier Undocking Presentation
Benefits
• Elimination of the continuous weight tracking effort results in significant labor savings, relieving Trades and Ship's Force of the burden of weekly weight status reporting.
• Carefully controlled flooding of the dry-dock with one or more pauses to correct impending list may add several hours to the actual undocking evolution, but is well worth maintaining control and safety of the ship.

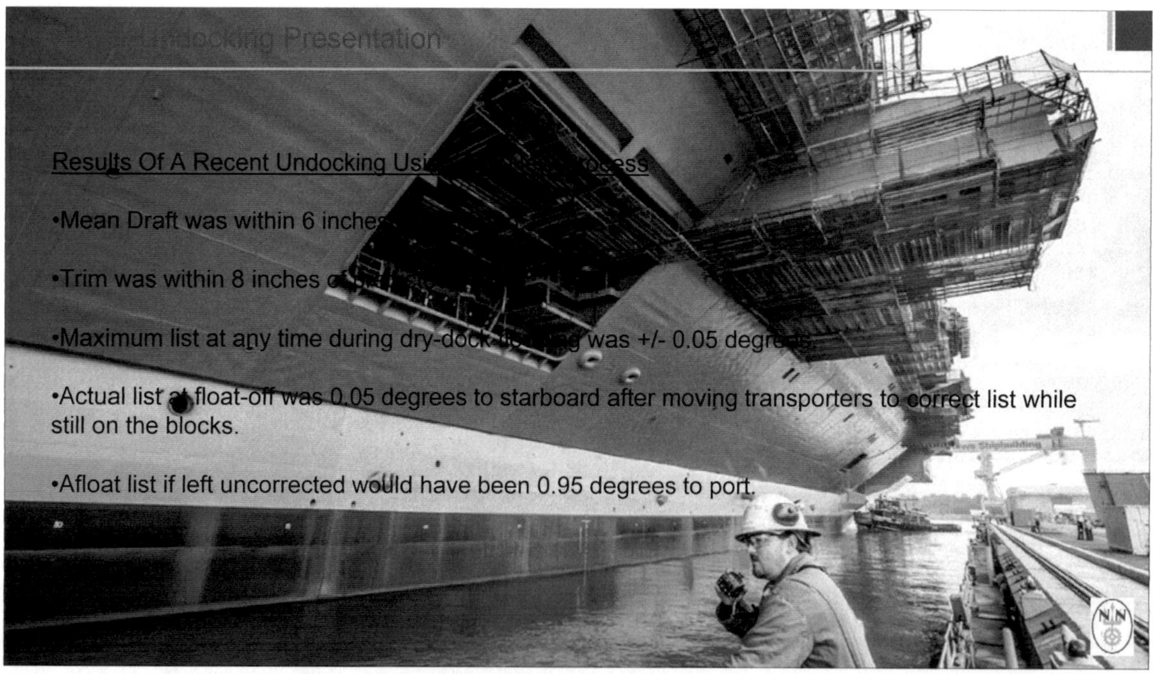
Undocking Presentation
Results Of A Recent Undocking Usi
•Mean Draft was within 6 inches
•Trim was within 8 inches
•Maximum list at any time during dry-dock was +/- 0.05 degrees.
•Actual list float-off was 0.05 degrees to starboard after moving transporters to correct list while still on the blocks.
•Afloat list if left uncorrected would have been 0.95 degrees to port.

Newport News
Shipbuilding
A Division of Huntington Ingalls Industries

THE DOCKING REPORT

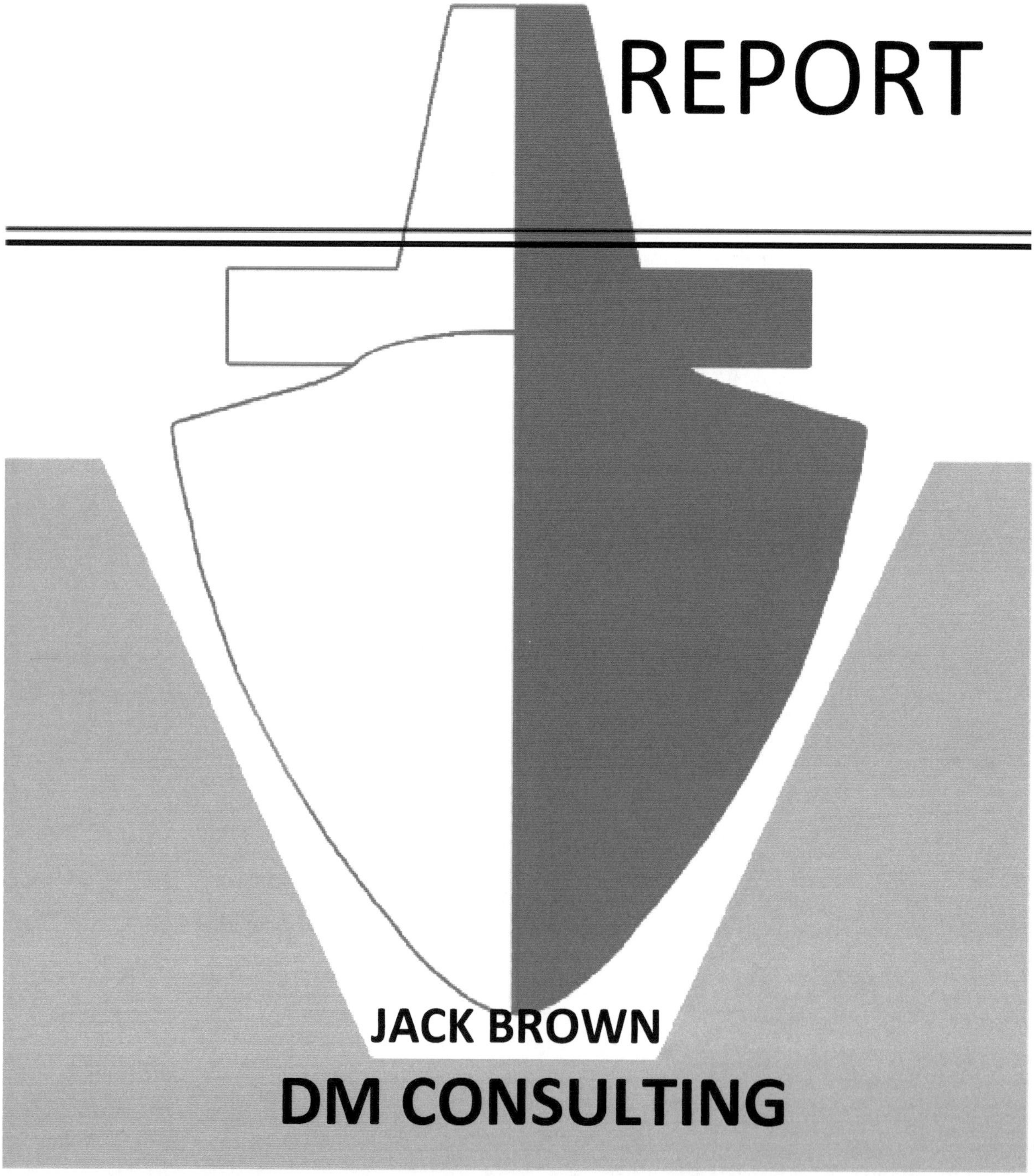

JACK BROWN

DM CONSULTING

THE DOCKING REPORT

Navy's also known as
NAVSEA REPORT 9997-1
(revised October 2015)
or
NAVSEA
S9086-7G-STM-010/CH-997
(replaces NAVSEA REPORT
9997/2 November 1977
Formerly NAVSHIPS 9070/2)

Available on the internet as well as the end of this presentation.

WHY?

PRE-DOCKING
and
POST-DOCKING
Record of Data

WHEN?

AS ARRIVED CONDITION
AS RELEASED CONDITION

WHAT IS COVERED?

1. ALL SHAFT DATA ON EACH SHAFT AS SET FORTH MUST BE TAKEN AND RECORDED ON PAGES ONE THROUGH FOUR OF REPORT FORM.

BY WHOM?

1. THE DOCKING FACILITY.

DOCKING REPORT

NAVSEA REPORT 9997-1
NAVSEA S9086-7G-STM-010/CH-997

FROM:				DATE
TO COMMANDING OFFICER, U.S.S				
SHIP NAME		HULL NUMBER:		
PLACE DOCKED	DATE DOCKED	DATE UNDOCKED	POSITION NUMBER THIS DOCKING	
			DIST. S.R.P. TO AFTMOST KEEL BLOCK	
DOCKING PLAN NUMBER	PLAN NUMBERS FOR ZINCS OR CATHODIC PROTECTION		REASON FOR DOCKING	

– BEFORE ENTERING DATA INTO THIS FORM, READ THE INSTRUCTIONS ON PAGES 8 & 9 –

SHAFTS & BEARINGS (AS-ARRIVED CONDITIONS)

SHAFT (DESIGN PARAMETERS)		
DESIGN DIAMETER	DESIGN CLEARANCE:	MAXIMUM ALLOWANCE CLEARANCE:

CONDITION OF SHAFTING AND MATERIAL	
NO 1 OR STBD OUTBD	
NO 2 OR STBD INBD	
NO 3 OR PORT INBD	
NO 4 OR PORT OUTBD	

RECORD SHAFT SECTION SERIAL NUMBERS, SHAFT-TO-STAVE RETAINER MEASUREMENT (STSR) AND SHAFT SLEEVE DIAMETER									
	STERN TUBE			INTERMEDIATE STRUT			MAIN STRUT		
READINGS TAKEN BY	SERIAL NO.	STSR – AFT END OF AFT BEARING	SHAFT SLEEVE DIA	SERIAL NO	STSR – AFT END	SHAFT SLEEVE DIA	SERIAL NO.	STSR – FWD END	SHAFT SLEEVE DIA
NO 1 OR STBD OUTBD									
NO 2 OR STBD INBD									
NO 3 OR PORT INBD									
NO 4 OR PORT OUTBD									

RECORD AVERAGE BEARING LAND INTERNAL DIAMETER (ID), AVERAGE BEARING BUSHING OUTER DIAMETER (OD) AND THE CLEARANCE BETWEEN THEM (BEARING LAND ID – BEARING BUSHING OD)			
NO 1 OR STBD OUTBD			
LOCATION	BEARING LAND ID	BEARING BUSHING OD	CLEARANCE
FWD STERN TUBE BEARING			
1. (FWD BEARING LAND)			
2.			
3.			
4. (AFT BEARING LAND)			
AFT STERN TUBE BEARING			
1. (FWD BEARING LAND)			
2.			
3.			
4. (AFT BEARING LAND)			
INTERMEDIATE BEARING			
1. (FWD BEARING LAND)			
2.			
3.			
4. (AFT BEARING LAND)			
MAIN STRUT BEARING			
1. (FWD BEARING LAND)			
2.			
3.			
4.			
5.			
6.			
7. (AFT BEARING LAND)			
NO 2 OR STBD INBD			
LOCATION	BEARING LAND ID	BEARING BUSHING OD	CLEARANCE
FWD STERN TUBE BEARING			

NAVSEA 9997/1 (Rev. OCT 2015)

DOCKING REPORT

1. (FWD BEARING LAND)			
2.			
3.			
4. (AFT BEARING LAND)			
AFT STERN TUBE BEARING			
1. (FWD BEARING LAND)			
2.			
3.			
4. (AFT BEARING LAND)			
INTERMEDIATE BEARING			
1. (FWD BEARING LAND)			
2.			
3.			
4. (AFT BEARING LAND)			
MAIN STRUT BEARING			
1. (FWD BEARING LAND)			
2.			
3.			
4.			
5.			
6.			
7. (AFT BEARING LAND)			
NO 3 OR PORT INBD			
LOCATION	BEARING LAND ID	BEARING BUSHING OD	CLEARANCE
FWD STERN TUBE BEARING			
1. (FWD BEARING LAND)			
2.			
3.			
4. (AFT BEARING LAND)			
AFT STERN TUBE BEARING			
1. (FWD BEARING LAND)			
2.			
3.			
4. (AFT BEARING LAND)			
INTERMEDIATE BEARING			
1. (FWD BEARING LAND)			
2.			
3.			
4. (AFT BEARING LAND)			
MAIN STRUT BEARING			
1. (FWD BEARING LAND)			
2.			
3.			
4.			
5.			
6.			
7. (AFT BEARING LAND)			
NO 4 OR PORT OUTBD			
LOCATION	BEARING LAND ID	BEARING BUSHING OD	CLEARANCE
FWD STERN TUBE BEARING			
1. (FWD BEARING LAND)			
2.			
3.			
4. (AFT BEARING LAND)			
AFT STERN TUBE BEARING			
1. (FWD BEARING LAND)			
2.			
3.			
4. (AFT BEARING LAND)			
INTERMEDIATE BEARING			
1. (FWD BEARING LAND)			
2.			
3.			
4. (AFT BEARING LAND)			
MAIN STRUT BEARING			
1. (FWD BEARING LAND)			
2.			
3.			
4.			
5.			
6.			
7. (AFT BEARING LAND)			

NAVSEA 9997/1 (Rev. OCT 2015)

STAVE & BEARING READINGS
***NOTE: N/A THE LINES FOR SYSTEMS WITH FEWER BEARINGS THAN LISTED.**

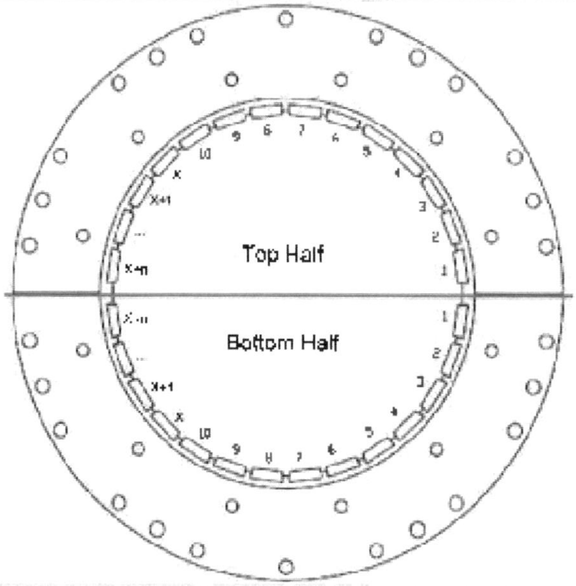

BEARING CLEARANCES AT EACH INDIVIDUAL STAVE

NO 1 OR STBD OUTBD

	FWD STERN TUBE	AFT STERN TUBE	FWD INTERMEDIATE STRUT	AFT INTERMEDIATE STRUT	FWD MAIN STRUT	AFT MAIN STRUT
			TOP HALF			
1						
2						
3						
4						
5						
6						
7						
8						
9						
10						
11						
12						
13						
14						
			BOTTOM HALF			
1						
2						
3						
4						
5						
6						
7						
8						
9						
10						
11						
12						
13						
14						

NO 2 OR STBD INBD

	FWD STERN TUBE	AFT STERN TUBE	FWD INTERMEDIATE STRUT	AFT INTERMEDIATE STRUT	FWD MAIN STRUT	AFT MAIN STRUT
			TOP HALF			
1						
2						
3						
4						
5						

6						
7						
8						
9						
10						
11						
12						
13						
14						
			BOTTOM HALF			
1						
2						
3						
4						
5						
6						
7						
8						
9						
10						
11						
12						
13						
14						

NO 3 OR PORT INBD

	FWD STERN TUBE	AFT STERN TUBE	FWD INTERMEDIATE STRUT	AFT INTERMEDIATE STRUT	FWD MAIN STRUT	AFT MAIN STRUT
			TOP HALF			
1						
2						
3						
4						
5						
6						
7						
8						
9						
10						
11						
12						
13						
14						
			BOTTOM HALF			
1						
2						
3						
4						
5						
6						
7						
8						
9						
10						
11						
12						
13						
14						

NO 4 OR PORT OUTBD

	FWD STERN TUBE	AFT STERN TUBE	FWD INTERMEDIATE STRUT	AFT INTERMEDIATE STRUT	FWD MAIN STRUT	AFT MAIN STRUT
			TOP HALF			
1						
2						
3						
4						
5						
6						
7						
8						
9						
10						
11						
12						
13						
14						
			BOTTOM HALF			
1						
2						
3						
4						

NAVSEA 9997/1 (Rev. OCT 2016)

2. ALL PROPELLER DATA ON EACH PROPELLER MUST BE TAKEN AND RECORDED ON PAGE FIVE OF THE REPORT.

3. ALL RUDDER AND DIVING PLANE DATA MUST BE TAKEN AND ALSO RECORDED ON PAGE FIVE OF THE REPORT.

5						
6						
7						
8						
9						
10						
11						
12						
13						
14						

PROPELLERS (AS-ARRIVED CONDITIONS)

*ATTACH RESULTS OF VISUAL TECHNICAL INSPECTION (VTI)

PROPELLERS	INBD OUTBD		
INBD. DIA.	PITCH	INB'D PLAN NO. AND REVISION	NUMBER OF BLADES
OUTBD. DIA.	PITCH	OUTB'D	MATERIAL

RUDDERS AND DIVING PLANES

RUDDERS AND DIVING PLANES (DESIGN PARAMETERS)	
POST DESIGN DIAMETER IN.	DESIGN BEARING CLEARANCE (UPPER/LOWER) IN.
BEARING MANUFACTURER	

RUDDERS AND DIVING PLANES (AS-ARRIVED CONDITIONS)			
POST DIAMETER	PORT ACTUAL IN.	STARBOARD ACTUAL IN.	CENTERLINE ACTUAL IN.
BEARING CLEARANCE	PORT (UPPER/LOWER) IN.	STARBOARD (UPPER/LOWER) IN.	CENTERLINE (UPPER/LOWER) IN.
IF RUDDER REMOVED REPORT PERCENT CONTACT OF BLUE CHECK ON TAPER	PORT %	STARBOARD %	CENTERLINE %
CONDITION OF RUDDERS/DIVING PLANES			
CONDITION OF BEARINGS (GREASED, RUST PRESENT, ETC)			

RUDDERS AND DIVING PLANES (AS-RELEASED CONDITIONS)			
POST DIAMETER (N/A IF NO WORK ACCOMPLISHED)	PORT IN.	STARBOARD IN.	CENTERLINE IN.
BEARING CLEARANCE	PORT (UPPER/LOWER) IN.	STARBOARD (UPPER/LOWER) IN.	CENTERLINE (UPPER/LOWER) IN.
IF RUDDER REMOVED, REPORT PERCENT CONTACT OF BLUE CHECK ON TAPER	PORT %	STARBOARD %	CENTERLINE %
INFLATABLE BOOT REPLACED (N/A IF SHIP DOES NOT HAVE INFLATABLE BOOT) YES ☐ NO ☐ N/A ☐			

4. ALL SONAR SYSTEMS MUST BE INSPECTED, MEASURED AND DATA RECORDED ON PAGE SIX OF THE REPORT.

SONAR		
SONAR REMOVED THIS DOCKING		
DOME TYPE	SERIAL	
SONAR EXISTING AT UNDOCKING DOME TYPE	SERIAL	WORK DONE – INCLUDE PAINTING

REMARKS *(Continue on separate sheet, if necessary)*

5. PAGE SEVEN OF THE REPORT ADDRESSES THE APPLICATION OF HULL PRESERVATIVES, TO INCLUDE ALL USED COATINGS, MIL-SPECS, AND MANUFACTURE OF THE COATINGS, AND OTHER REQUIRED DATA.

SELECT HULL & APPENDAGES COATING APPLICATIONS ACCOMPLISHED THIS DOCKING
– N/A TO SUBMARINES –

NAVSEA SI 009-32 FY ' ' APPLICABLE								
			COATINGS DATA					
LOCATION	a.) Table # b.) Line #	SURFACE PREPARATION	# COATS	MIL-SPEC	MANUFACTURER & TRADE NAME	100% RECOAT	TOUCH-UP AND OVER-COAT	SPOT TOUCH-UP ONLY
UNDERWATER HULL (Keel to top of boottop. Including propulsion shaft outboard bearing voids)	a.) b.)							
RUDDERS AND STRUTS (If different than underwater hull coating system above)	a.) b.)							
SEA CHESTS (List by Hull Opening ## in this block.)	a.) b.)							
FIN STABILIZERS (Not Applicable to Aircraft Carriers)	a.) b.)							
DIELECTRIC SHIELDS	a.) b.)							
EXPOSED AREAS OF OUTBOARD SHAFTING COVERED BY GRP	a.) b.)							
EXTERIOR SURFACES ABOVE BOOTTOP (Freeboard surfaces only)	a.) b.)							
OTHER: For surface preparation and / or coating applications which differ from those above, for systems not listed above, or for more room for systems listed above, enter the data below or use similar format to report the data on a supplementary page.								
OTHER	a.) b.)							
OTHER	a.) b.)							
OTHER	a.) b.)							

NAVSEA 9997/1 (Rev. OCT 2015)

TO ASSIST WITH THE TIMELY COMPLETION OF THE REPORT, PAGES EIGHT AND NINE ARE INTENDED TO PROVIDE PRECISE AND TO THE POINT DIRECTIONS FOR COMPLETING THE REPORT.

INSTRUCTIONS

GENERAL:

1. **From**: Enter reporting activity.
2. **Date**: Enter date of report.
3. To: Commanding Officer of the ship drydocked.
4. **Ship Name**: Name of ship drydocked
5. **Hull Number**: Hull number of ship drydocked
6. **Place Docked**: Naval Shipyard, station, industrial activity drydock number.
7. **Docking Date**: Enter date ship drydocked.
8. **Undocking Date**: Enter date ship is undocked.
9. **Position No.**: Enter position number employed when ship is drydocked.
10. **Dist.**: S.R.P. to aftmost keel block: Enter distance from Stern Reference Point.
11. **Docking Plan No.**: List docking plan number and revision employed to drydock the ship.
12. **Plan Numbers for Zinc or Cathodic Protection**: Indicate applicable plan number for installation.
13. **Reason for Docking**: State major reason for drydocking. i.e., propeller repairs, sonar dome and hoist replacement, hull damage, regular overhaul, etc.

SHAFTS & BEARINGS (AS-ARRIVED CONDITION):

1. Enter shaft design diameter, design clearance and maximum allowed clearance from information provided by ship, manufacturers instructions and Naval Ships Technical Manual
2. Record as-arrived condition of shafting covering system. Report exposed metal surfaces, condition (i.e., good condition, cracks, scored, corroded, etc) and adherence.
3. Record shaft serial number, STSR measurement and shaft sleeve diameter. Enter name of personnel taking readings.
   Note: STSR (shaft-to-stave retainer) measurement is the distance between the bearing stave retainer ring and the shaft sleeve at the 6 o'clock position. See Naval Ships Technical Manual S9086-HN-STM-010, Chapter 244 or Underwater Ship Husbandry Manual S0600-AA-PRO-160 Chapter 16 for additional details.
4. Record average bearing land inner-diameter, average bearing bushing outer-diameter and the clearance at each land location.
5. Record as-arrived bearing clearances at each individual stave.

PROPELLERS (AS-ARRIVED CONDITION):

1. Attach results of as-arrived Visual Technical Inspection (VTI).
2. Enter propeller data for each propeller as determined by examination of hub stamping. List the propeller material as follows:
   M - Manganese Bronze
   N - Nickel Aluminum Bronze
   A - Manganese Nickel Aluminum Bronze
   S - Stainless Steel
   P - Special (if special materials other than above are used, note under remarks)

RUDDER AND DIVING PLANE DATA (AS-ARRIVED & AS-RELEASED CONDITIONS):

1. Enter post design diameter, design bearing clearance (upper and lower) and bearing manufacturer.
2. Record Post Diameter and bearing Clearance for each rudder and diving plane as determined by measurement at time of docking for as-arrived conditions. Take such rudder bearing clearances as are practicable without unshipping the rudder. Use R for rudder and D for diving planes. Use F and A to designate forward and aft where necessary. Divide port and starboard columns where necessary. Check rudder nut lock.
3. Record as-arrived condition of rudders. Report coating deterioration, structural defects and overall condition. i.e. good condition, cracks, scored, corroded, etc.
4. Record as-arrived condition of rudder bearings. Report if the bearings have been greased, if rust is present, etc.
5. Record Post Diameter and bearing Clearances for each rudder and diving plane as determined by measurement at time of undocking for as-released conditions (if any repairs/replacements have been made). Take such rudder bearing clearances as are practicable without unshipping the rudder. Use R for rudder and D for diving planes. Use F and A to designate forward and aft where necessary. Divide port and starboard columns where necessary. Check rudder nut lock.
6. Record percent contact of blue check on taper (if rudders are removed).
7. Record whether the inflatable boot has been replaced. Check "Yes" or "No", leave blank if ship does not have inflatable boot.

SONAR: Report dome type and serial number. The serial appears on the skirt and in the shipping orders.

REMARKS: Note when extensive repair or replacement of plates, seams, rivets, etc., is accomplished, including data concerning locations, actions and description of conditions which necessitated actions. Include any other information considered desirable for reference purposes, such as special block heights or other departures from docking plan. Attach additional sheets if required.

COATING APPLICATIONS ACCOMPLISHED (AS-RELEASED):

1. Record NAVSEA Standard Item 009-32 fiscal year (in Table title).
2. Record Table and Line numbers of Standard Item 009-32 followed for preservation.
3. Record NACE/SSPC method of preparation (SSPC-SP 1, SP 3, SP 11, SP 10, etc.) in sufficient detail; for example, SSPC-SP 1. "detergent-wash", or "solvent-wash", or "water-wash".
4. Record number of coats for each type of coating (1, 2, etc.).
5. Record MIL-SPEC for each type of coating.
6. Record Manufacture & Trade Name(s) for each coating (both primer/anticorrosive and antifouling)
7. Record the scope of the preservation work accomplished as follows:
   - For 100% Recoat, place an "X" in the appropriate block.
   - For Touch up and Overcoat, estimate zone worked by recording frame numbers.
   - For Spot touch up Only, note spot repair areas by recording frame numbers.

***SEE NEXT PAGE FOR CONTINUATION OF INSTRUCTIONS FOR SUPPLEMENTARY FORMS** (to be filled out in addition to 9997/1, if criteria listed for each form is met)*

INSTRUCTIONS - CONTINUED

SUPPLEMENTARY FORMS (to be filled out in addition to 9997/1, if criteria listed for each form is met):

1. **Propellers:** Submit a supplementary form NAVSEA 9997/2 PROPELLER REPORT and/or NAVSEA 9997/3 PROPELLER REPORT – DETACHABLE BLADE DATA ONLY-if repairs are accomplished, or propellers or blades are replaced.
2. **Shafts, Bearings, Fairwaters, Ropeguards and Stern Tube Seals:** Submit supplementary form NAVSEA 9997/4, ONLY when repairs or replacements are made or when corrosion or wear is extensive or severe.
   a. Use one form sheet for each line of shafting.
   b. The sketch should be modified by crossing out excess bearings for systems with fewer bearings than sketched.
   c. Clearance readings: identification of material; condition of material; and work accomplished on Bearings, Fairwaters, Ropeguards, Journals, Shaft Covering, Packing, and Stern Tube Seals should be indicated in each applicable box provided for the recording of the condition and work accomplished by entering the proper symbols. If the symbol "O" (Other) is used to indicate information, an Arabic number may be used, such as "2", to refer to an explanatory footnote "2".
   d. Record shaft serial number, STSR measurement, and shaft sleeve diameter. Enter name of personnel taking readings
   e. Record average bearing land inner-diameter, average bearing bushing outer-diameter and the clearance at each land location.
   f. Record As-Released Shaft-to- Stave Readings (bearing clearances at each individual stave). List size (oversized/undersized) of stave, whether the rubber or backing was over/undersized, manufacturer used (Duramax, etc.) and material for each stave. Enter appropriate symbol for type of bearing material in each box
      W – Wood
      PP – Phenolic Plastic
      R – Rubber
      F - Fiberglass
      O – Other (specify type)
   g. List rotating coupling cover repairs and o-ring information.
   h. Record Mil-Spec for shaft coating system applied.

SHIP CLASS SPECIFIC SUPPLEMENTARY FORMS:

AIRCRAFT CARRIERS ONLY:

1. **Rudders:** Submit supplementary form NAVSEA 9997/6, for all aircraft carrier dockings, for as-arrived and as-released Rudder Bearing Measurements and Conditions.
2. **Underwater Hull & Appendages Fairing Compound:** Submit supplementary form NAVSEA 9997/7, ONLY when Fairing Compound Repairs are accomplished. Record Underwater Hull & Appendages Fairing Compound Repairs and list type and amount for each of the locations.

DIFFICULTIES AND MAJOR PROBLEMS I HAVE FOUND IN THE PAST, IN PREPARING THIS REPORT IN A TIMELY MANNER IS THE TASK OF GATHERING ALL RELATIVE INFORMATION FROM ALL WHO HAVE INPUTS FOR THE REPORT.

THE MORE INVOLVED INDIVIDUALS FOR THE PRIMARY CONTRACTOR AND ANY NUMBER OF SUB-CONTRACTORS WHO GATHER DATA AS WELL AS ALL THE ITEMS WITH REGARD TO PAINT AND HULL COATING CAN BECOME A TRUE TASK, OR EVEN AN ORDEAL.

THE FOLLOWING NINE PAGES PROVIDE A COMPLETE COPY OF THE MOST CURRENT DOCKING REPORT AND ARE SUITABLE FOR REPRODUCTION.

<u>THE EASY PART IS OVER NOW.</u>

HOW DO WE ADDRESS THE NEED FOR A REPORT CONCERNING NON NAVY VESSELS DOCKING IN NON NAVY FACILITIES.

THIS NEED CAN BE ADDRESSED BY FIRST DETERMINING WHAT IS NEEDED TO MAKE A FULL AND MEANINGFUL

The Drydocking Report documents the information relative to the Drydocking. The report at a minimum should include the following:

A. Docker Master
B. Dry Dock used
C. Name of vessel
D. Type of vessel and class
E. Owner's name
F. Shipyards Drydocking Plan
    i. Reference number
    ii. Docking position utilized
    iii. Location of document
G. Height of the keel blocks above Drydock floor
H. Principal dimensions of the vessel: (LOA, LBP., Beam (extreme and molded))

I. Vessel drydocking condition
    i. Drydocking drafts
    ii. Drydocking displacement
    iii. Time vessel landed on blocks
    iv. Height of tide at time of drydocking

J. Vessel undocking condition
    i. Undocking drafts
    ii. Undocking displacement
    iii. Time vessel floated off the blocks
    iv. Height of tide at time of undocking (if applicable)

K. Date and time vessel entered the drydock

L. Date and time vessel exited the drydock

M. Vessel work accomplished (hull painting, alternations, etc.) which will impact future drydockings

N. Photos of blocks (keel, side, shores) and ship (before and after landing on the blocks)

QUESTIONS AND CONCERNS

Should a standard form be generated to be used by commercial docking facilities?

By whom? Could this fall under the guidance being developed in the "Safe Operation and Maintenance of DRY DOCK FACILITIES" (ASCE Manuals and Reports on Engineering Practive No. 121) Docking Reports are mentioned several times in the aforementioned manual.

FROM:				DATE
TO COMMANDING OFFICER, U.S.S.				
SHIP NAME			HULL NUMBER:	
PLACE DOCKED	DATE DOCKED	DATE UNDOCKED	POSITION NUMBER THIS DOCKING DIST. S.R.P. TO AFTMOST KEEL BLOCK	
DOCKING PLAN NUMBER	PLAN NUMBERS FOR ZINCS OR CATHODIC PROTECTION			REASON FOR DOCKING

– BEFORE ENTERING DATA INTO THIS FORM, READ THE INSTRUCTIONS ON PAGES 8 & 9 –

SHAFTS & BEARINGS (AS-ARRIVED CONDITIONS)

SHAFT (DESIGN PARAMETERS)		
DESIGN DIAMETER:	DESIGN CLEARANCE:	MAXIMUM ALLOWANCE CLEARANCE:

CONDITION OF SHAFTING AND MATERIAL	
NO 1 OR STBD OUTBD	
NO 2 OR STBD INBD	
NO 3 OR PORT INBD	
NO 4 OR PORT OUTBD	

RECORD SHAFT SECTION SERIAL NUMBERS, SHAFT-TO-STAVE RETAINER MEASUREMENT (STSR) AND SHAFT SLEEVE DIAMETER									
	STERN TUBE			INTERMEDIATE STRUT			MAIN STRUT		
READINGS TAKEN BY	SERIAL NO.	STSR – AFT END OF AFT BEARING	SHAFT SLEEVE DIA	SERIAL NO	STSR – AFT END	SHAFT SLEEVE DIA	SERIAL NO.	STSR – FWD END	SHAFT SLEEVE DIA
NO 1 OR STBD OUTBD									
NO 2 OR STBD INBD									
NO 3 OR PORT INBD									
NO 4 OR PORT OUTBD									

RECORD AVERAGE BEARING LAND INTERNAL DIAMETER (ID), AVERAGE BEARING BUSHING OUTER DIAMETER (OD) AND THE CLEARANCE BETWEEN THEM (BEARING LAND ID – BEARING BUSHING OD)			
NO 1 OR STBD OUTBD			
LOCATION	BEARING LAND ID	BEARING BUSHING OD	CLEARANCE
FWD STERN TUBE BEARING			
1. (FWD BEARING LAND)			
2.			
3.			
4. (AFT BEARING LAND)			
AFT STERN TUBE BEARING			
1. (FWD BEARING LAND)			
2.			
3.			
4. (AFT BEARING LAND)			
INTERMEDIATE BEARING			
1. (FWD BEARING LAND)			
2.			
3.			
4. (AFT BEARING LAND)			
MAIN STRUT BEARING			
1. (FWD BEARING LAND)			
2.			
3.			
4.			
5.			
6.			
7. (AFT BEARING LAND)			
NO 2 OR STBD INBD			
LOCATION	BEARING LAND ID	BEARING BUSHING OD	CLEARANCE
FWD STERN TUBE BEARING			

NAVSEA 9997/1 (Rev. OCT 2015)

1. (FWD BEARING LAND)			
2.			
3.			
4. (AFT BEARING LAND)			
AFT STERN TUBE BEARING			
1. (FWD BEARING LAND)			
2.			
3.			
4. (AFT BEARING LAND)			
INTERMEDIATE BEARING			
1. (FWD BEARING LAND)			
2.			
3.			
4. (AFT BEARING LAND)			
MAIN STRUT BEARING			
1. (FWD BEARING LAND)			
2.			
3.			
4.			
5.			
6.			
7. (AFT BEARING LAND)			
NO 3 OR PORT INBD			
LOCATION	BEARING LAND ID	BEARING BUSHING OD	CLEARANCE
FWD STERN TUBE BEARING			
1. (FWD BEARING LAND)			
2.			
3.			
4. (AFT BEARING LAND)			
AFT STERN TUBE BEARING			
1. (FWD BEARING LAND)			
2.			
3.			
4. (AFT BEARING LAND)			
INTERMEDIATE BEARING			
1. (FWD BEARING LAND)			
2.			
3.			
4. (AFT BEARING LAND)			
MAIN STRUT BEARING			
1. (FWD BEARING LAND)			
2.			
3.			
4.			
5.			
6.			
7. (AFT BEARING LAND)			
NO 4 OR PORT OUTBD			
LOCATION	BEARING LAND ID	BEARING BUSHING OD	CLEARANCE
FWD STERN TUBE BEARING			
1. (FWD BEARING LAND)			
2.			
3.			
4. (AFT BEARING LAND)			
AFT STERN TUBE BEARING			
1. (FWD BEARING LAND)			
2.			
3.			
4. (AFT BEARING LAND)			
INTERMEDIATE BEARING			
1. (FWD BEARING LAND)			
2.			
3.			
4. (AFT BEARING LAND)			
MAIN STRUT BEARING			
1. (FWD BEARING LAND)			
2.			
3.			
4.			
5.			
6.			
7. (AFT BEARING LAND)			

NAVSEA 9997/1 (Rev. OCT 2015)

STAVE & BEARING READINGS
***NOTE: N/A THE LINES FOR SYSTEMS WITH FEWER BEARINGS THAN LISTED.**

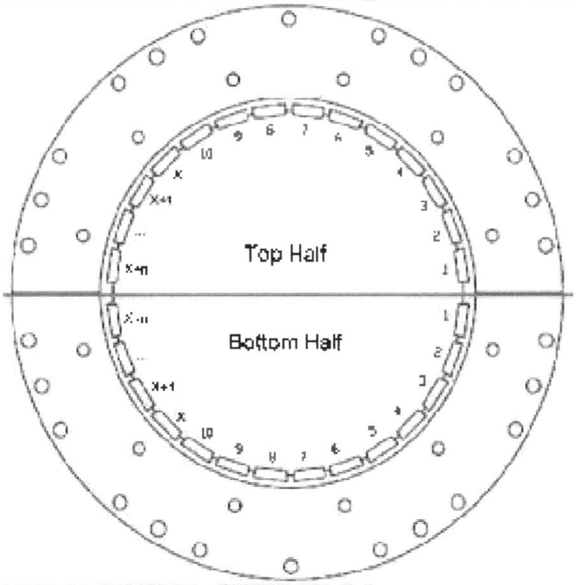

BEARING CLEARANCES AT EACH INDIVIDUAL STAVE

	NO 1 OR STBD OUTBD					
	FWD STERN TUBE	AFT STERN TUBE	FWD INTERMEDIATE STRUT	AFT INTERMEDIATE STRUT	FWD MAIN STRUT	AFT MAIN STRUT
	TOP HALF					
1						
2						
3						
4						
5						
6						
7						
8						
9						
10						
11						
12						
13						
14						
	BOTTOM HALF					
1						
2						
3						
4						
5						
6						
7						
8						
9						
10						
11						
12						
13						
14						
	NO 2 OR STBD INBD					
	FWD STERN TUBE	AFT STERN TUBE	FWD INTERMEDIATE STRUT	AFT INTERMEDIATE STRUT	FWD MAIN STRUT	AFT MAIN STRUT
	TOP HALF					
1						
2						
3						
4						
5						

NAVSEA 9997/1 (Rev. OCT 2015)

6						
7						
8						
9						
10						
11						
12						
13						
14						
			BOTTOM HALF			
1						
2						
3						
4						
5						
6						
7						
8						
9						
10						
11						
12						
13						
14						

NO 3 OR PORT INBD

	FWD STERN TUBE	AFT STERN TUBE	FWD INTERMEDIATE STRUT	AFT INTERMEDIATE STRUT	FWD MAIN STRUT	AFT MAIN STRUT
			TOP HALF			
1						
2						
3						
4						
5						
6						
7						
8						
9						
10						
11						
12						
13						
14						
			BOTTOM HALF			
1						
2						
3						
4						
5						
6						
7						
8						
9						
10						
11						
12						
13						
14						

NO 4 OR PORT OUTBD

	FWD STERN TUBE	AFT STERN TUBE	FWD INTERMEDIATE STRUT	AFT INTERMEDIATE STRUT	FWD MAIN STRUT	AFT MAIN STRUT
			TOP HALF			
1						
2						
3						
4						
5						
6						
7						
8						
9						
10						
11						
12						
13						
14						
			BOTTOM HALF			
1						
2						
3						
4						

NAVSEA 9997/1 (Rev. OCT 2015)

5						
6						
7						
8						
9						
10						
11						
12						
13						
14						

PROPELLERS (AS-ARRIVED CONDITIONS)

*ATTACH RESULTS OF VISUAL TECHNICAL INSPECTION (VTI)

PROPELLERS	INBD OUTBD		
INBD. DIA.	PITCH	INB'D PLAN NO. AND REVISION	NUMBER OF BLADES
OUTBD. DIA.	PITCH	OUTB'D	MATERIAL

RUDDERS AND DIVING PLANES

RUDDERS AND DIVING PLANES (DESIGN PARAMETERS)	
POST DESIGN DIAMETER IN.	DESIGN BEARING CLEARANCE (UPPER/LOWER) IN.
BEARING MANUFACTURER	

RUDDERS AND DIVING PLANES (AS-ARRIVED CONDITIONS)			
POST DIAMETER	PORT ACTUAL IN.	STARBOARD ACTUAL IN.	CENTERLINE ACTUAL IN.
BEARING CLEARANCE	PORT (UPPER/LOWER) IN.	STARBOARD (UPPER/LOWER) IN.	CENTERLINE (UPPER/LOWER) IN.
IF RUDDER REMOVED REPORT PERCENT CONTACT OF BLUE CHECK ON TAPER	PORT %	STARBOARD %	CENTERLINE %
CONDITION OF RUDDERS/DIVING PLANES			
CONDITION OF BEARINGS (GREASED, RUST PRESENT, ETC)			

RUDDERS AND DIVING PLANES (AS-RELEASED CONDITIONS)			
POST DIAMETER (N/A IF NO WORK ACCOMPLISHED)	PORT IN.	STARBOARD IN.	CENTERLINE IN.
BEARING CLEARANCE	PORT (UPPER/LOWER) IN.	STARBOARD (UPPER/LOWER) IN.	CENTERLINE (UPPER/LOWER) IN.
IF RUDDER REMOVED, REPORT PERCENT CONTACT OF BLUE CHECK ON TAPER	PORT %	STARBOARD %	CENTERLINE %
INFLATABLE BOOT REPLACED (N/A IF SHIP DOES NOT HAVE INFLATABLE BOOT) YES ☐ NO ☐ N/A ☐			

SONAR		
SONAR REMOVED THIS DOCKING		
DOME TYPE	SERIAL	
SONAR EXISTING AT UNDOCKING DOME TYPE	SERIAL	WORK DONE – INCLUDE PAINTING

REMARKS *(Continue on separate sheet, if necessary)*

SELECT HULL & APPENDAGES COATING APPLICATIONS ACCOMPLISHED THIS DOCKING
– N/A TO SUBMARINES –

NAVSEA SI 009-32 FY ' ' APPLICABLE									
LOCATION	a.) Table # b.) Line #	SURFACE PREPARATION	COATINGS DATA			100% RECOAT	TOUCHUP AND OVER-COAT	SPOT TOUCHUP ONLY
			# COATS	MIL-SPEC	MANUFACTURER & TRADE NAME			
UNDERWATER HULL (Keel to top of boottop, including propulsion shaft outboard bearing voids)	a.) b.)							
RUDDERS AND STRUTS (If different than underwater hull coating system above)	a.) b.)							
SEA CHESTS (List by Hull Opening '##' in this block.)	a.) b.)							
FIN STABILIZERS (Not Applicable to Aircraft Carriers)	a.) b.)							
DIELECTRIC SHIELDS	a.) b.)							
EXPOSED AREAS OF OUTBOARD SHAFTING COVERED BY GRP	a.) b.)							
EXTERIOR SURFACES ABOVE BOOTTOP (Freeboard surfaces only)	a.) b.)							
OTHER: For surface preparation and / or coating applications which differ from those above, for systems not listed above, or for more room for systems listed above, enter the data below or use similar format to report the data on a supplementary page.								
OTHER	a.) b.)							
OTHER	a.) b.)							
OTHER	a.) b.)							

NAVSEA 9997/1 (Rev. OCT 2015)

INSTRUCTIONS

GENERAL:

1. **From:** Enter reporting activity.
2. **Date:** Enter date of report.
3. To: Commanding Officer of the ship drydocked.
4. **Ship Name:** Name of ship drydocked
5. **Hull Number:** Hull number of ship drydocked
6. **Place Docked:** Naval Shipyard, station, industrial activity drydock number.
7. **Docking Date:** Enter date ship drydocked.
8. **Undocking Date:** Enter date ship is undocked.
9. **Position No.:** Enter position number employed when ship is drydocked.
10. **Dist.:** S.R.P. to aftmost keel block: Enter distance from Stern Reference Point.
11. **Docking Plan No.:** List docking plan number and revision employed to drydock the ship.
12. **Plan Numbers for Zinc or Cathodic Protection:** Indicate applicable plan number for installation.
13. **Reason for Docking:** State major reason for drydocking, i.e., propeller repairs, sonar dome and hoist replacement, hull damage, regular overhaul, etc.

SHAFTS & BEARINGS (AS-ARRIVED CONDITION):

1. Enter shaft design diameter, design clearance and maximum allowed clearance from information provided by ship, manufacturers instructions and Naval Ships Technical Manual
2. Record as-arrived condition of shafting covering system. Report exposed metal surfaces, condition (i.e., good condition, cracks, scored, corroded, etc) and adherence.
3. Record shaft serial number, STSR measurement and shaft sleeve diameter. Enter name of personnel taking readings.
   Note: STSR (shaft-to-stave retainer) measurement is the distance between the bearing stave retainer ring and the shaft sleeve at the 6 o'clock position. See Naval Ships Technical Manual S9086-HN-STM-010, Chapter 244 or Underwater Ship Husbandry Manual S0600-AA-PRO-160 Chapter 16 for additional details.
4. Record average bearing land inner-diameter, average bearing bushing outer-diameter and the clearance at each land location.
5. Record as-arrived bearing clearances at each individual stave.

PROPELLERS (AS-ARRIVED CONDITION):

1. Attach results of as-arrived Visual Technical Inspection (VTI).
2. Enter propeller data for each propeller as determined by examination of hub stamping. List the propeller material as follows:
   M - Manganese Bronze
   N - Nickel Aluminum Bronze
   A - Manganese Nickel Aluminum Bronze
   S - Stainless Steel
   P - Special (If special materials other than above are used, note under remarks)

RUDDER AND DIVING PLANE DATA (AS-ARRIVED & AS-RELEASED CONDITIONS):

1. Enter post design diameter, design bearing clearance (upper and lower) and bearing manufacturer.
2. Record Post Diameter and bearing Clearance for each rudder and diving plane as determined by measurement at time of docking for as-arrived conditions. Take such rudder bearing clearances as are practicable without unshipping the rudder. Use R for rudder and D for diving planes. Use F and A to designate forward and aft where necessary. Divide port and starboard columns where necessary. Check rudder nut lock.
3. Record as-arrived condition of rudders. Report coating deterioration, structural defects and overall condition, i.e. good condition, cracks, scored, corroded, etc.
4. Record as-arrived condition of rudder bearings. Report if the bearings have been greased, if rust is present, etc.
5. Record Post Diameter and bearing Clearances for each rudder and diving plane as determined by measurement at time of undocking for as-released conditions (if any repairs/replacements have been made). Take such rudder bearing clearances as are practicable without unshipping the rudder. Use R for rudder and D for diving planes. Use F and A to designate forward and aft where necessary. Divide port and starboard columns where necessary. Check rudder nut lock.
6. Record percent contact of blue check on taper (if rudders are removed).
7. Record whether the inflatable boot has been replaced. Check "Yes" or "No", leave blank if ship does not have inflatable boot.

SONAR: Report dome type and serial number. The serial appears on the skirt and in the shipping orders.

REMARKS: Note when extensive repair or replacement of plates, seams, rivets, etc., is accomplished, including data concerning locations, actions and description of conditions which necessitated actions. Include any other information considered desirable for reference purposes, such as special block heights or other departures from docking plan. Attach additional sheets if required.

COATING APPLICATIONS ACCOMPLISHED (AS-RELEASED):

1. Record NAVSEA Standard Item 009 32 fiscal year (in Table title).
2. Record Table and Line numbers of Standard Item 009-32 followed for preservation.
3. Record NACE/SSPC method of preparation (SSPC-SP 1, SP 3, SP 11, SP 10, etc.) in sufficient detail; for example, SSPC-SP 1 "detergent-wash", or "solvent-wash", or "water-wash".
4. Record number of coats for each type of coating (1, 2, etc.).
5. Record MIL-SPEC for each type of coating.
6. Record Manufacture & Trade Name(s) for each coating (both primer/anticorrosive and antifouling)
7. Record the scope of the preservation work accomplished as follows:
   - For 100% Recoat, place an "X" in the appropriate block.
   - For Touch up and Overcoat, estimate zone worked by recording frame numbers.
   - For Spot touch up Only, note spot repair areas by recording frame numbers.

SEE NEXT PAGE FOR CONTINUATION OF INSTRUCTIONS FOR SUPPLEMENTARY FORMS (to be filled out in addition to 9997/1, if criteria listed for each form is met)

INSTRUCTIONS - CONTINUED

SUPPLEMENTARY FORMS (to be filled out in addition to 9997/1, if criteria listed for each form is met):

1 **Propellers**: Submit a supplementary form NAVSEA 9997/2 PROPELLER REPORT and/or NAVSEA 9997/3 PROPELLER REPORT – DETACHABLE BLADE DATA ONLY-if repairs are accomplished, or propellers or blades are replaced.

2 **Shafts, Bearings, Fairwaters, Ropeguards and Stern Tube Seals**: Submit supplementary form NAVSEA 9997/4, ONLY when repairs or replacements are made or when corrosion or wear is extensive or severe.

- a. Use one form sheet for each line of shafting.
- b. The sketch should be modified by crossing out excess bearings for systems with fewer bearings than sketched.
- c. Clearance readings: identification of material; condition of material; and work accomplished on Bearings, Fairwaters, Ropeguards, Journals, Shaft Covering, Packing, and Stern Tube Seals should be indicated in each applicable box provided for the recording of the condition and work accomplished by entering the proper symbols. If the symbol "O" (Other) is used to indicate information, an Arabic number may be used, such as "2", to refer to an explanatory footnote "2".
- d. Record shaft serial number, STGR measurement, and shaft sleeve diameter. Enter name of personnel taking readings
- e. Record average bearing land inner-diameter, average bearing bushing outer-diameter and the clearance at each land location.
- f. Record As-Released Shaft-to- Stave Readings (bearing clearances at each individual stave). List size (oversized/undersized) of stave, whether the rubber or backing was over/undersized, manufacturer used (Duramax, etc.) and material for each stave. Enter appropriate symbol for type of bearing material in each box
  W – Wood
  PP – Phenolic Plastic
  R – Rubber
  F - Fiberglass
  O – Other (specify type)
- g. List rotating coupling cover repairs and o-ring information.
- h. Record Mil-Spec for shaft coating system applied.

SHIP CLASS SPECIFIC SUPPLEMENTARY FORMS:

AIRCRAFT CARRIERS ONLY:

1 **Rudders**: Submit supplementary form NAVSEA 9997/6, for all aircraft carrier dockings, for as-arrived and as-released Rudder Bearing Measurements and Conditions.

2 **Underwater Hull & Appendages Fairing Compound**: Submit supplementary form NAVSEA 9997/7, ONLY when Fairing Compound Repairs are accomplished. Record Underwater Hull & Appendages Fairing Compound Repairs and list type and amount for each of the locations.

The 9th Dry Dock Conference and Advanced Training Forum
Houston, TX, USA
5-6 May 2017

LAUNCHING AND DRYDOCKING OF NEW HOBART CLASS DESTROYERS

N A Clark, Naval Architect, ASC Shipbuilding, AUS Nicholas.Clark@asc.com.au

A M Eriksson, Naval Architect, ASC Shipbuilding, AUS Anna.Eriksson@asc.com.au

ABSTRACT

This paper discusses the establishment and implementation of safe launch and drydocking processes at the recently established ASC Shipbuilding shipyard in South Australia. It will discuss the steps taken in preparing the launch of the first ship built at the shipyard through to the launch of the second ship and also the first dry docking undertaken in the yard.

ASC Shipbuilding was commissioned for the purpose of building surface ships for the Royal Australian Navy (RAN). The shipyard was constructed adjacent to the existing Submarine yard that has built and maintained the Collins Class Submarines. Between the two yards lies the Common User Facility (CUF), which was constructed by Defence SA, a SA state owned company. Defence SA facilities include a wharf, dryberth for ship consolidation and dockings, shiplift, shore connection services, and land for future expansion. Aside from conducting commercial dockings, the Defence SA site is (currently) predominantly used for constructing and launching Australia's new Hobart Class Air Warfare Destroyers (AWD).

The first ship launched in the ASC Shipbuilding yard was NUSHIP Hobart on the 23rd of May 2015. This was significant as it was a first of class ship; first ship built by the yard, and first build by a new organization under an "Alliance Arrangement". The Air Warfare Destroyer Alliance is responsible for delivering three Hobart Class Air Warfare Destroyers and their support systems to Defence. The Alliance is made up of shipbuilder ASC, mission systems integrator Raytheon Australia and the Government's Capability Acquisition and Sustainment Group, supported by shipbuilder manager, Navantia.

What was particularly challenging was that, commencing 18 months prior to launch, all shipyard procedures and documentation needed to be created from scratch, using a first principles approach and advice from various capability partners and technical support network in industry. The rigorous preparations resulted in a successful launch via shiplift on schedule, with no leaks, predicted trim achieved, no damage to the ship and a slight heel to starboard. Lessons learnt from the Hobart's launch process were implemented in the preparation and launch of NUSHIP Brisbane, some 18 months later on the 15th of December 2016. Improvements included reducing the mass and Centre of Gravity (CoG) prediction time, from 9 weeks to 10 days, consolidating requirements and minimizing documentation, clarifying responsibilities, and reducing the number of stakeholders required to a core group of 6-8 people. The launch was again successful, with no leaks, target trim achieved, heel just over 1 degree and no damage to the ship.

One week after the launch of Brisbane, Hobart was docked for the first time in preparation for acceptance trials in January 2017. This was another significant evolution as first for class and yard. The processes established during the preceding two launches were largely adopted for the docking evolution, which meant only 4 weeks was required to define work scope, roles and responsibilities. In addition to experience gained from launches, procedures established over 18 months of construction alongside the wharf for managing watertight integrity and weight management were used to control changes to the ship in dock. The result was a safe docking and undocking for the ship and personnel, on schedule, with no damage to the hull or appendages and no change to heel or trim when floating off the blocks, compared to entry to the dock.

In conducting the discussed evolutions, the significance of preparing and undertaking launching and dockings for the first time in a new shipyard and a new organisation was evident.

The shipyard has demonstrated success in undertaking preparations and managing risk and safety to achieve on time launching and docking evolutions. After the first launch, significant efficiency gains in subsequent evolutions were a result of established processes combined with experience and lessons learnt from previous evolutions including clearly defined and understood roles, responsibilities, and technical requirements.

NOMENCLATURE

+ve	Positive
ABR	Australian Book of Reference (Navy)
AWD	Air Warfare Destroyer
BIW	Bath Iron Works
BL	Baseline
COG	Centre of Gravity (m)
CUF	Common User Facility
DM	Dockmaster
FWD	Forward
GM	Metacentric Height (m)
GRP	Glass Reinforced Polymer
KED	Key Event Definition
LAT	Lowest Astronomical Tide
LCG	Longitudinal Centre of Gravity (m)
LOA	Length Overall (m)
MS	Amidships
NUSHIP	New Ship/Submarine (Yet to be commissioned)
OQE	Objective Quality Evidence
OST	Outfitting Support Tower
RAN	Royal Australian Navy
SA	South Australia
STBD	Starboard
SWBS	Ship Weight Breakdown Structure
TCG	Transverse Centre of Gravity (m)
TDP	Technical Data Package
TTS	Total Transfer Systems
VCG	Vertical Centre of Gravity (m)
WT	Watertight

1. INTRODUCTION

This paper is about the preparation required to launch the new Hobart Class Destroyers, for the first time, in a new shipyard by a new organization. The Hobart class is a class of three new multi-purpose air defence destroyers of 147 m LOA, with full load displacement of 6300 metric tonnes. The ships are transferred from the 315 m long dryberth consolidation site, via a TTS transfer system with greater than 10,000 t capacity, and launched and docked at the Defence SA Shiplift. The shiplift is 156x34m, currently certified by Navy to 9300 t with a greater than 13600 t total capacity, and can take a ship with draft of up to 9.5m.

At launch, each ship is in majority constructed and ready for in water activation. Each ship is not expected to be docked again until immediately prior to acceptance trials.

2. NUSHIP HOBART LAUNCH

The first ship launched in the ASC Shipbuilding yard was NUSHIP Hobart on the 23rd of May 2015. This was significant as it was a first of class ship, first ship built by the yard, and first build by a new organization under an "Alliance Arrangement". In addition, the shiplift and dry berth transfer system were owned and managed by Defence SA, a third party to the shipyard, not what could be considered a normal arrangement for a shipyard.

What was particularly challenging was that, commencing 18 months prior to launch, all shipyard procedures and documentation needed to be created from scratch, using a first principles approach and advice from various colleagues in industry. Additionally, as with all first milestones for ship 1, the float off received a significant amount of scrutiny which added to the workload e.g. one week long alliance review, two week long risk workshops, 10 revisions to the draft Key Event Document to satisfy stakeholders, and regular attendances of 30-40 people at weekly float off meetings.

Each review would generate actions that would require response and closure, with responsibility to address engineering actions lying with the Naval Architect who was also responsible for their own scope of work in preparation for launch in addition to collecting/reviewing evidence of completed work against launch requirements. Education of some stakeholders, which did not have backgrounds in shipbuilding, was also required to set expectations. New processes were also required to be established for in-water management of watertightness, load condition and stability, which did not exist. It was a busy time.

2.1. DEVELOPMENT OF REQUIREMENTS

As Hobart's launch was the first launch for the shipyard, no formal procedures existed for preparing the ship, facilities, and carrying out the launch, or 'float off'. Rather than try to re-invent the wheel, contact was made in January 2014 to various industry participants and colleagues to gather existing rules, procedures, guidelines and best practices. Sources included and thanks goes to:

- Bernie Dobson, Defence SA,
- Jeff Neale, Bath Iron Works (BIW),
- Joe Stiglich, DM Consulting,
- Begona Rodriguez, Navantia,
- The Naval Architecture team at Naval Tech Bureau Hydrodynamics Cell (formerly Directorate of Naval Platform Systems),
- Lieutenant. Aaron Gill Royal Australian Navy (RAN),
- The Naval Architecture team at ASC North, and
- David Gonzalez, ASC Shipbuilding.

Documents varied from generic procedures for the preparation and undocking of ships, shipyard specific procedures, procedures specific to the parent design F-100 series, and regulatory rules and guidelines for undocking of Naval Ships in Australia. As most of these documents were in a checklist format, the first draft of the master checklist was formed using parts from all documents that were relevant to the type of ship, shipyard, facilities and local rules and regulations.

After several review cycles to satisfy all stakeholders, the first revision of the checklist was completed. This would form the 'Annex A – Master Sequence', the primary source of technical requirements for float off. The Annexure was to a primary document called the Float off Key Event Definition (KED). The KED contained generic requirements about the intent of launch, the overall scope, scheduling information and safety information. Additionally, there were further secondary checklists that were referenced by the Master Sequence in Annex A.

Annex A was divided into several sections, with two for preparing the ship prior to rolling away on the hardstand: Hardware preparation (physical completion, checks and testing of the ship) and Documentation and Planning. The remaining sections were checklists for Transfer to Shiplift (from the dry berth consolidation location), Float Off (including watertight integrity test and ceremonial float off), and berthing at the wharf. Each requirement step had person/s assigned as responsible parties to carry out the task and where required provide evidence, and at completion sign off the activity.

Requirements within Annex A included watertight boundary testing and required evidence, V-Line upper limits to which main bulkheads shall be made watertight (two consecutive compartment flooding scenario), first and second level valves from the hull to be proven watertight and tagged out closed, valves required to vent sea chests and bow thruster cavity be locked open to match design hydrostatics for trim purposes.

2.2. IMPLEMENTATION OF REQUIREMENTS

As mentioned previously, a unique aspect of the Shipyard is that the consolidation and transfer area and shiplift are owned and controlled by Defence SA, a state owned company. Defence SA and the AWD Alliance/ASC Shipbuilding have a contractual agreement for priority access to space on the CUF, and use of the shiplift. As a result, the Dockmaster is not directly employed by the company, and that every time the ship is moved, either via transfer system, shiplift or tug, there is a transfer of material control and responsibility.

Each party had a checklist of requirements that must be satisfied prior to handing over/accepting control of the vessel during any of the above mentioned evolutions. For ASC Shipbuilding, the checklists developed in the KED were used to provide assurance to the shipyard signatory that all checks/plans/process/documentation had been carried out, completed and recorded. Any gaps in Objective Quality Evidence (OQE) prior to handover were risk assessed, with justification to proceed or mitigation applied if required, and approved by the appropriate level of authority commensurate with the risk.

2.3. WEIGHT ESTIMATE FOR FLOAT OFF

A mass displacement and Centre of Gravity (COG) prediction was required 3 months prior to launch, in order to support the publication of a launching document by the designer. This was a challenge as the ship changes considerably in the last 3 months prior to launch, no material tracking system existed to provide a status of what was on/off the ship, and it was the first time such a prediction had been made. The total time taken for this activity was 3 months full time for one person.

As a starting point, it was decided to use the lightship weight report, to the 5 digit SWBS level, as a baseline for the launch prediction. A combination of sources was used to obtain a status of build completion against the lightship report, as well as deadweight items and temporary works.

2.3.1. LIGHTSHIP

The ship design is broken down to ~2200 drawings at the 5 digit SWBS level. For the purpose of the weight report that summarises the lightship, they are separated to a further total of ~3500 unique lines items

Each 5 digit drawing is broken down into at least one or a series of work orders, which are issued to production. The planning department had a reporting tool that output the status of all work orders issued and future planned.

By checking the planned completion status of all work orders, the completion status as of the end of April (for May 23rd launch) was determined. If all work orders containing weight (install orders e.g.) for a given drawing were complete by April, they were counted as being 100% complete.

The check of drawings containing weight was not arbitrary, and carried out on a case by case basis, with adjustments between 0% and 100% used where necessary based on work order and area manager feedback.

Work scopes that fall outside of normal planning work orders such as accommodation and furniture loadout, as well as deck screed (both sub contracted works), were assessed for levels of completion based on area manager feedback and tracking tools, and % completion assigned to the relevant 5 digit TDP drawings.

2.3.2. TEMPORARY WORKS

All ship areas were surveyed for temporary works by the Naval Architecture team in conjunction with Production Area Managers. The focus of the survey was to identify any temporary works that were planned to remain at launch, as well as to give a high level status of lightship completion. This survey yielded many observations, the main ones of which included:

1. Exterior scaffold would be removed for launch, with internal scaffold expected to remain in blocks containing intake and uptake spaces to be completed. Each scaffold group by block was surveyed by engineering and support services (with centres of gravity assigned), and in total came to 50.5 tonnes.
2. Temporary wooden false floors would remain, particularly in the combat compartments in lieu of GRP floors. Flooring centres of gravity were accurately determined from material mass properties and from 5 digit TDP false floor drawings.
3. Furniture, the scope of which mainly lies outside the normal work order system (it is a contracted scope of work), were loaded out into final compartments below main deck prior to access cuts being closed.

2.3.3. LIGHTSHIP FLUIDS AND TEST EQUIPMENT

The planned status of Lightship fluids in equipment, and test equipment planned to be left onboard for launch, was obtained in advance. Any weight type that could potentially change between prediction and launch day were checked several times in the lead up to launch.

2.3.4. USE OF LAUNCH PREDICTION

The mass and COG prediction were used to determine the ballasting arrangement for float off. As the ship is stable by design, even in an incomplete lightship condition, ballast was primarily to obtain the desired trim. At the Dockmaster and designers recommendation and from first principles, a slight aft trim of +10-12 cm by the stern was required.

2.3.5. VALIDATION OF PREDICTION PRIOR TO LAUNCH

During the ship transfer from dry berth to shiplift, the transfer system and shiplift were used as the first validation checks of the prediction, to check for overall displacement and LCG position. See Table 2-1 below for the accuracy of each system, as stated by Defence SA.

Table 2-1 - TTS and Shiplift Accuracy

Accuracy	System TTS	System Shiplift
(%)	+/- 3	+/- 1
% @5000T (T)	+/- 150	+/- 50
Distance (mm)	+/-1000	+/-100

Both systems were deemed accurate enough to check for LCG, with an uncertainty range of +/-2cm in trim at launch displacement. Accuracy was not deemed sufficient for TCG checks, with the uncertainty in heel prediction due to error yielding a range of +/- 2.5degrees. Certainly, if the measurement exceeded this range, there would be a significant error in the launch ballast case that would have required rectifying.

At the first instance of lifting the ship on the hydraulic transfer system, Defence SA had a hold point to check mass, with an acceptable range from the prediction of +5% -2%. As per results shown in Table 2-2, the accuracy of the prediction was within the error limits of the transfer system.

Table 2-2 – Transfer System Weighing Results - Hobart

Description	Predicted Mass (T)	Measured Mass (T)
Ship at Launch	5048.0	
Launch Cradle	475.0	
Cradle (Aft STBD)	14.6	
Cradle (Fwd STBD)	11.9	
Topside scaffold	5.0	
Upper Limit (+5%)	5830	
TOTAL	5552	**5472**
Lower Limit (-2%)	5441	

Figure 2-1 - Hobart at the end of side transfer, prior to forward transfer to shiplift.

2.4. WATERTIGHT INTEGRITY TEST

Due to the advantage of launching from a shiplift, prior to the official launch day, the ship was lowered until 10% of the ships mass displacement load remained on the blocks, to conduct a ship wide watertight integrity test. Each major subdivision compartment was manned and rounds were conducted as the ship was lowered, and held at the test depth for several hours. Tide was not a factor in restricting the duration of this test, as the Defence SA shiplift pocket is sufficiently deep that the Hobart Class can clear the blocks at Full Load Displacement during the Lowest Astronomical Tide (LAT).

In the event that significant defects were found, there were two spare days prior to launch day to rectify any defects, and conduct another WT integrity test.

The result was that out of all the submerged watertight boundaries, pipes, valves, only one 1st boundary valve had a leaking flange, which was slow enough to be fixed in-water by re-applying the torque sequence. Overall it was a resounding success for the shipyard, and the ship was now ready for the official launch.

2.5. RESULTS

On the 23rd of May 2015, NUSHIP Hobart floated off the blocks for the first time and on schedule. The trim target of 10 cm by the stern was achieved with a result of 9 cm, and heel was 1.6 degrees to starboard. The heel was larger than expected, however nothing more could have been done to detect the transverse CoG, as the distance from the centreline was in the order of 10's of millimetres.

Following the watertight integrity test, no further leaks occurred during and after float off.

After the float off ceremony, Hobart was hauled out of the shiplift via Capstans and tugs, and transferred to the wharf. Once at the wharf, the ship was in a lockdown state for 24 hours in case of any leaks emerging, none of which occurred. During the 24 hour period, ballast was transferred to correct the heel, and was used as a basic 'builders inclining' to confirm the ships GM. This information was then used for subsequent in-water management calculations until the full inclining experiment was carried out.

See Table 2-3 below, for a comparison of launch displacement prediction and CoG results, obtained from draft marks and the builders inclining conducted immediately post launch.

Table 2-3 - Comparison of Hobart Launch results to prediction

Ballasted Condition	MASS (T)	VCG (m) (+ve Over BL)	LCG (m) (+ve FWD MS)	TCG (m) (+VE STBD)
Prediction	5086	8.307	-1.730	-0.001
Actual	5000	8.175	-1.512	0.041

2.6. LESSONS LEARNT AND RECOMMENDATIONS

It is normal that on this type of project, there are many lessons learnt from the first of class that are applied to following ships. In the case of float off, lessons learnt and recommendations include:

1. The watertight requirements were validated due to no incidents occurring.
2. Reduce number of stakeholders at meetings from 30+, to core members who are responsible e.g. Production, Scheduling, Engineering.
3. Revise float off requirements to reduce non value adding steps, and consolidate requirements to improve clarity.
4. Reduce the body of the Key Event Definition document to the bare minimum.
5. Assign task owners to requirement steps earlier, to force responsible people to action their work in a timely manner. For Hobart, engagement with customer's witnesses was often too late. Early intervention for ship 2 will remedy this.
6. Revise and make clear Objective Quality Evidence (OQE) requirements, to ensure all documentation is captured first time to minimize rework and risk assessments.
7. Decompress the ship transfer to shiplift schedule. This is always a compromise between schedule demand and mitigating schedule risk.
8. Procure all temporary equipment required for launch well enough in advance, so as to not put the launch at risk.
9. Ballast as many tanks as allowed at construction tower, prior to moving to shiplift to save time at the shiplift.
10. Minimize the number of risk and safety reviews, where practical.
11. Make all documentation generic for the launch of this class of destroyer, to minimize re-work for ship 3 launch and subsequent dock/undock evolutions.

3. NUSHIP BRISBANE LAUNCH

Following the launch of NUSHIP Hobart on May 23rd 2015, NUSHIP Brisbane's launch was scheduled for December 15th 2016, 18 months later. Due to other commitments in preparing and conducting Hobarts inclining experiment, inclining report, trim and stability book, and preparation for Builders Sea Trials, there was significantly reduced time to prepare for Brisbane's launch.

3.1. REVISION OF REQUIREMENTS

The focus of revising launch requirements was to consolidate steps so no scope overlap existed, make the document and requirements generic for any ship in the class, replace responsible person names with position titles, update witness and OQE requirements based on customer and engineering requirements, and reduce the number of stakeholders required to manage float off preparations to a minimum.

With a December launch in mind, requirements were revised with the Launch Project Manager and Ship Managers input, over a two month period leading up to June 2016. Draft requirements were released August 3rd (Formalised in early November) for production to schedule to, and begin notifying commonwealth witnesses of witness and hold points. At this point, production began collecting OQE against requirements, and notifying Customer Representatives of upcoming witness hold points.

Following NUSHIP Hobarts Inclining and Builders Sea Trials, Brisbane Float off meetings and preparation recommenced in October 2016, three months prior to launch.

3.2. IMPLEMENTATION OF REQUIREMENTS

Noting lessons learnt from Hobart with regards to over-populated meetings, the float off Project Manager ensured that meetings were attended by one key person from each department, up to 6 in total. Each department representative was responsible for reporting the status of their assigned tasks that directly related to requirements in the KED Annex A. A dedicated scheduler managed the schedule view of activities, and tracked progress accordingly.

Another improvement over Hobart was early engagement with the customer and making clear to all parties what OQE, witness and hold point requirements were to be followed. For many tasks, the customer required witness and in some cases hold points, as well as requiring the receipt of Objective Quality Evidence (OQE). In another improvement over Hobart, the customer helped minimised this bottleneck, by adding two more Design Acceptance Representatives (DAR's) early in the process, who were pro-active in highlighting witness and hold points, and ensuring that they were available for as many as possible.

Overall, compared to Hobart's launch, gathering of OQE and witnessing of work occurred earlier and in a steady manner, when compared to Hobarts bow wave of work at the end. Production was prepared well in advance, and the customer was well engaged, with some room for improvement still remaining. There was very little clarification required on requirements, which was proof that the revision work carried out had been successful in articulating to the workforce what was required.

3.3. WEIGHT ESTIMATE

The launch weight prediction method for Brisbane was an evolution of what was developed for Hobart. Since the Hobart float off, refinements of the method were made to support the ships inclining experiment. Two iterations were conducted for Inclining, and an independent validation proved that the prediction method is very accurate. The sum of improvements meant that for Brisbane's displacement prediction, time was reduced significantly from 3 months full time, to 10 days full time. This was due to a combination of factors, the main one of which was continual efficiency improvements made in the processing of the data. Secondly, production had implemented a number of basic tracking systems on outfit items such as deck coverings, insulation, furniture installation etc. which reduced the time taken to calculate partially complete work scope. Other time reducing factors included further completion of ship scope, leaving less incomplete weights to check, and better housekeeping in production to minimize temporary works survey time.

From a time management point of view, the reduced time required to predict the launch displacement meant that the task could fit into the tight timeframe allowed, given conflicting priorities with NUSHIP Hobart and the submission deadline for the information to be 1 month prior to launch, rather than 3 months in advance compared to Hobart (as per contract).

At the first ship move hold point, the following results were obtained in Table 3-1.

Table 3-1 - Transfer System Weighing Results - Brisbane

Description	Predicted Mass (T)	Measured Mass (T)
Ship at Launch	4965.0	
Launch Cradle	475.0	
Upper Limit (+5%)	5712	
TOTAL	5440	5550
Lower Limit (-2%)	5331	

Figure 3-1 - Brisbane at the end of side transfer, prior to forward transfer to shiplift.

3.4. WATERTIGHT INTEGRITY TEST

Using the same approach as ship 1, Hobart, a watertight integrity test, whereby the ship is lowered until 10% mass displacement load is remaining on the keel blocks, was used to check for leaks. As per ship 1, one 1st boundary valve flange had a leak, which to be safe was repaired with the ship raised back up to the transfer level. A further final dip revealed no further leaks and proof Brisbane was ready for launch.

3.5. RESULTS

On the 15th of December 2016, NUSHIP Brisbane floated off the blocks for the first time and on schedule. The trim target of 11.5 cm by the stern was achieved with an actual trim of 13 cm, and heel was -1.1 degrees to Port. The heel was less than that for Hobart, whilst still leaving room for improvement at the behest of the Naval Architects pride!

Following the watertight integrity test, no further leaks occurred during and after float off. The ship was transferred to the wharf to be adjacent to NUSHIP Hobart, where the minor heel was corrected.

Overall, the ship was 40 tonnes heavier in lightship completion compared to Hobart. Whilst it was not directly measured, the reduced effort to the get to the point of float off was a significant cost saving for the company.

See Table 3-2 below, for a comparison of launch displacement prediction and CoG results, obtained from draft marks post launch. Compared to the launch of Hobart (Table 2-3), improvement was made in calculating both LCG and TCG positions.

Brisbane can be seen in Figure 3-2, with propellers shining in the sun, prior to launch.

Table 3-2 – Comparison of Brisbane Launch results to prediction

Ballasted Condition	MASS (T)	VCG (m) (+ve Over BL)	LCG (m) (+ve FWD MS)	TCG (m) (+VE STBD)
Prediction	5037	8.307	-1.720	0.000
Actual	5055	--------	-1.746	-0.020

Figure 3-2 - NUSHIP Brisbane, prior to Launch

3.6. LESSONS LEARNT

Much improvement was made between float offs of Hobart and Brisbane, with implemented lessons learnt leaving less room for improvement when looking ahead to the launch of the final ship in class, Sydney. Nevertheless, lessons learnt and potential improvements include:

1. Defence SA were able to move the ship from Outfit Support Tower (OST) to the shiplift in one day, which could save a day in the schedule for ship 3 if planned for (Brisbane had two days allocated to the OST-Shiplift). It should be noted however, that there was a spare day in the case of inclement weather or equipment issues.
2. Ensuring all parties who are responsible for providing assurance that testing and closure of watertight valves are complete, are reminded of the requirement to track work and produce evidence that the work has been completed.

4. HOBART DOCKING AND UNDOCKING

A docking of NUSHIP Hobart had been discussed for a longer period of time. Late in 2016 a decision was made to conduct a short, 14 day long docking over the Christmas period. The purpose was to undertake a confined scope of work in preparation for acceptance trials in January 2017.

The docking of Hobart took place only one week after the float-off of Brisbane. As with the float-off evolution, this was a significant event being the first ship of the class docked in the shipyard.

4.1. DOCKING ARRANGEMENT

A docking arrangement, different to the construction cradle, had been developed for the Defence SA common user facility shiplift prior to Hobart float-off in 2015. The docking arrangement was developed based on the Hobart Class ship designer Navantia's docking arrangement and adapted by Defence SA for trestles and shiplift frame spacing applicable to the Defence SA shiplift. The blocks, example shown in Figure 4-1, were manufactured on site by Defence SA based on a 3D hull model generated by an ASC Naval Architect. The docking arrangement differed significantly from the docking arrangement defined in the designer's technical data package. Supported by Defence SA and ASC Shipbuilding, Navantia carried out a substantial review, including Finite Element Analysis, to assess the ship's structural integrity on the suggested arrangement.

Figure 4-1 - Docking bilge blocks

4.2. REQUIREMENTS AND PROCESSES

For the Hobart docking evolution, the processes established during the preceding two launches were largely adopted. This meant that only 4 weeks were required to define work scope, roles and responsibilities.

A requirements document was developed based on the Float Off – Key Event Definition, DEF(AUST)5000–Vol 02 Pt 20, Slipping and Docking [1], and ABR 5225 Vol 1 Chap 5, External Maintenance Availabilities [2]. The requirements were tailored for the anticipated work scope to be covered for dockings during the construction and activation phases of the AWDs. As with the Float Off Key Event Definition, this document defined tasks and responsibilities in relation to the docking. It covered documentation and planning, technical requirements, hardware & ship preparation, personnel requirements, risk assessments and damage control.

Ship and Hardware preparations involved everything required to set the ship in the right configuration and condition for the docking, e.g. floating condition (heel and trim), tank states, rudder and propeller positions, systems status and valve configurations on tank boundaries and primary watertight boundaries, and the securing of movable equipment. It also defined requirements for preparing the docking cradle and any support services required for the evolution e.g. lighting and ventilation, ship access support and means of communications.

Due to most systems already being activated, requirements were specified for system status during docking, which related primarily to seawater systems. The reason was to avoid unknown weight changes during docking and undocking due to uncontrolled draining of systems, and to ensure systems would not be damaged during the time in the dock due to corrosion and internal growth.

The document also discussed data acquisition for naval architecture checks and calculations and included required checklists for ship condition, weight control and watertightness.

As with the preceding documents for the float-off evolutions, it defined the docking and undocking sequence and stated requirement for each of the discrete steps of the evolution. Each requirement was contained in a sign-off checklist where the responsible person was also defined. For those requirements associated with a potential catastrophic consequence if not fulfilled e.g. hull valve configuration and tank states, independent checks were carried out by engineering.

The focus for the documentation specific to this evolution, in comparison to previous evolutions though, was to define requirements on weight management and watertight integrity. For weight and watertight integrity management in the dock, the intent was to track changes to the ship during the docking period rather than doing a full ship survey prior to undocking. Therefore, there were strict requirements on how the work onboard was managed. A full work scope was defined prior to the docking and any changes to the scope had to go through the Docking Project Manager.

Since float-off of NUSHIP Hobart, processes had also been in place to manage the safety of the ship and personnel while being constructed and activated alongside the wharf. These processes involved, in particular, the In Water Stability Management Plan which defines how to manage watertight integrity and weight changes to the ship, and places strict limits on weight changes that have to be reported and assessed before initiated. It also uses the well-established shipyard work permit system to approve and control breaching of watertight boundaries (internal transversal bulkheads and shell).

Adopting these processes while in the dock allowed effective control of weight changes and changes to the watertight boundaries.

The weight limit set for reporting was 200 kg. In addition weight changes exceeding 200 kg also had to be recorded in a weight log. A recording requirement was also set in regards to breaching of watertight boundaries.

Figure 4-2 and Figure 4-3 show the ship entering the dock in preparation for docking.

Figure 4-2 - Ship entering the shiplift for docking

Figure 4-3 - Ship entering shiplift for docking – view from flight deck

4.3. DOCKING

As with the float-off evolutions there were handover points to the experienced team at Defence SA, led by Dockmaster Bernie Dobson, who would perform the actual docking and undocking.

The shipyard had beforehand defined a docking condition to achieve the desired floating condition for the ship. The displacement of the ship at the time was 5500 Tonnes and a slight aft trim was required. The target trim for the evolution was 7 cm stern down with 0 degrees heel.

When entering the shiplift, the ship had a slight heel that was adjusted using fresh water wing tanks. To achieve required trim there would normally be two options available, changes to fuel or ballast tanks.

With the cradle arrangement resulting in a stern overhang, and the surface on the hardstand being of concrete construction, overhang load needed to be minimized. While the overhang load introduced by having ballast tanks filled would be acceptable on the shiplift (due to the flexibility of the shiplift), they were required to be drained when transferring on to the concrete hardstand.

The intent was therefore to manage trim using the considerable amount of fuel that was held onboard the ship at the time, as the fuel tanks were within the longitudinal limits of the cradle. Unfortunately, only a few days in advance of the docking, repair work was required on the fuel system with system isolations being put in place. This made it impossible to transfer fuel forward to aft. With the fuel system inoperable, the only option was therefore to adjust trim with ballast.

This complicated the docking evolution due to the following reasons.

1. The only power available in the shiplift was ship's generators. No power supply or temporary lighting was available in the shiplift once generators were shut down for docking.
2. The method to drain ballast tanks while fully docked would be to use portable submersible pumps. This would require manual pumping by opening tank lids.
3. With poor lighting this would be hazardous to personnel with risk of potential damage to completed and inspected compartments.

Instead a decision was made to land the ship on the keel blocks, haul the bilge blocks and raise the ship until 10% of the displacement was resting on the blocks, approximately 500 tonnes. At this intermediate position, the aft ballast tank (20T) used for trim adjustment at landing on the blocks was pumped empty, diesel generators shut down and remaining hull valves were shut.

During docking, the load on each of the shiplift hoists was monitored and a validation of load predictions was carried out.

Figure 4-4 and Figure 4-5 show the ship while being raised on the shiplift during the docking.

Figure 4-4 - Ship landed and raised

Figure 4-5 - Ship partially docked

4.4. IN THE DOCK

While in the dock, with processes in place and the scope of work for the docking undertaken being well defined beforehand, the delta in watertight integrity and weight changes could be tracked using the established processes.

Work was carried out as per schedule and only complicated by a storm coming through, which damaged a substantial amount of care and protection put in place for deck blasting and painting.

Figure 4-6 and Figure 4-7 show the ship fully docked in the shiplift before transferring to the hardstand.

4.5. UNDOCKING

The undocking evolution went as per plan. After 12 days on the hardstand, the ship was transferred to the shiplift and undocked the following day.

Although there had been no announced changes to the ship's tank state, due to the potential catastrophic consequence of unknown major weight changes and/or hull boundaries not being watertight, checks were carried out of tank levels with all tanks being dipped prior to undock. Hull valves and overboard discharges on the primary watertight boundaries (hull shell) were also inspected without any work being undertaken on them.

Weight surveys were carried out before docking, and a review of the status of the ship was carried out prior to undocking to ensure there had been no significant changes to onboard weight.

Approximately a 20 m^3 difference was noted for the total volume in tanks as compared to docking (total of 400 m^3 onboard of which 300 m^3 was fuel). This difference was attributed to cumulative error in dipping tolerance. The resulting moment change due to noted differences in tank levels were assessed and was deemed to cause no significant impact on stability, trim or heel that would introduce any risk at undocking.

Coming out of the dock the ship was again held at an intermediate position with 10 % of the displacement resting on the blocks. Watertight integrity checks were carried out at the intermediate position with sentries in all below waterline compartments with hull penetrations. In this position, ship's generators were again started up and ballasting took place with an aim to achieve the target trim of 7 cm before floating off.

Time pressure was experienced during the float off phase, due to a tight schedule that included tug transfer to the wharf for berthing. During ballasting, a tank level sensor indicated that one of the aft starboard ballast tanks was being filled in addition to the centreline tank. This was later confirmed to be an issue with the tank sensor. It however required some investigation and caused a delay in the ballasting process. This meant that a new suing condition was calculated with a minimum trim required. The lowering of the ship continued while the centreline ballast tank was progressively filled during the last phase of the float-off.

Again, the expertise within Defence SA was very valuable in terms of performing the actual undocking of the ship.

Figure 4-6 - TTS trolley system lined up in preparation for ship transfer to runway

Figure 4-7 - Starboard blocking - NUSHIP Hobart fully docked in shiplift

Figure 4-9 - Example of bilge block fitting showing that which required the largest correction, which was quite minor. This block was at the forward most location, with greatest angle at contact.

4.6. LESSONS LEARNT

It was proven through the load distribution on the shiplift that some adjustments could be made to the docking arrangement post docking. This involved substituting the large aft most keel block with a standard size keel block, refer Figure 4-8. In general, the fitting of the bilge blocks were very good, with only a few blocks requiring adjustment, refer example in Figure 4-9.

With experience from this first docking evolution, the main requirements document can now be updated and simplified. Primarily, the number of steps requiring sign-off at each step in the docking sequence can be reduced. In addition, requirements on the status of hull penetrations will be revised and the scope reduced (deck drains e.g.).

The result using the established processes and the vast experience within Defence SA was a safe docking and undocking for the ship and personnel, on schedule, with no damage to the hull or appendages and no change to heel or trim coming off the blocks as compared to docking.

Figure 4-8 - Aft trestles with bracing

Figure 4-10 - NUSHIP Hobart fully docked in the shiplift

5. CONCLUSIONS

The shipyard has demonstrated success in undertaking preparations and managing risk and safety to achieve on time launching and docking evolutions.

Methods developed for launch condition weight estimates have also been proven to be very accurate, to within 1% of actual launch displacements.

In conducting the discussed evolutions, the significance of preparing and undertaking launching and dockings for the first time in a new shipyard and a new organisation has been evident.

Lessons learnt from the evolutions have primarily followed on from experience gained in the first evolution undertaken. The improvements in the processes have primarily involved reducing the number of people involved and technical reviews required, and the number of explicit low level requirements implemented in the initial master sequence.

It should be emphasised though, for the first evolution and being an inexperienced shipyard, that there was a need to have exhaustive, explicit and detailed requirements and well defined roles appointed to specific personnel in the organisation, to ensure tasks and responsibilities were clear with no part of the required work scope left out.

Other lessons learnt underlined the importance of early definition of task owners, task witnesses, commencement of tracking of work and collection of OQE. However, these again are improvements that come with experience.

With the experience in the organisation gained after the first evolution, a number of reviews, requirements and tasks were removed or made generic, and with the number of people reduced, significant efficiency gains were obtained in the subsequent evolutions.

6. ACKNOWLEDGEMENTS

In undertaking the work scope discussed in this paper many people were involved. We particularly want to thank Bernie Dobson, Paul Bates and their team at Defence SA for their considerable experience and support and our fellow colleagues at ASC Shipbuilding, who have worked tirelessly to prepare the ships for each and every launch and docking.

7. REFERENCES

1. Commonwealth of Australia, *Slipping and Docking, DEF(AUST)5000 Volume 2 Whole of Ship Requirements, Part 20, May 2003*

2. Commonwealth of Australia, *External Maintenance Availabilities, Australian Book of Reference 5225 Volume 1 Revision 5, Chapter 5,* March 2011.

MTS Conversion Docking

Norfolk Naval Shipyard
Portsmouth, VA

1

History: Norfolk Naval Shipyard

- Built: 1767
- Originally named Gossport Shipyard, it is the oldest industrial facility that belongs to the U.S. NAVY.
- Destroyed during Revolutionary War and Civil War.
- First docking in 1833.

2

History: Dry Dock No. 3

Built: 1911

Type: Graving with Pile Foundation

Dimensions:

700'L x 94'W x 27' D

Time to Flood:

2hrs

Dewatering Pumps:

190kgpm total capacity

Superflood Capable

3

Superflooding

Required to raise the water level in Dry Dock No. 2 & 3 for vessels with a deep draft.

Super flood is done by back flooding the main chamber from the dewatering flume through the main chamber flooding valve..

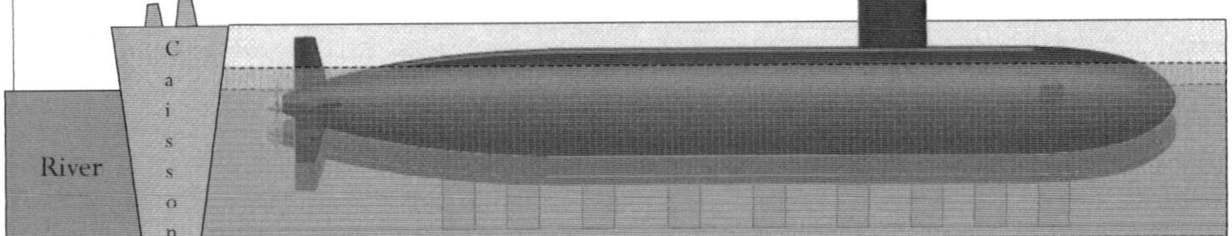

MTS Project Overview

Converting a U.S. NAVY submarine into a Moored Training Ship (MTS):

- 32 month availability
- 3 new hull sections added
- 1 hull section removed

5

Strongback Build

Non-standard NNSY build to accommodate for special availability

6

Docking

Special docking requiring specific tides and Superflooding to accommodate.

7

Segmenting

Separating hull sections so that new ones can be added.

8

Lining up SPMT's

Lifting and relocating Hull Sections.

9

Parallel parking a sub.

10

Lift new sections into the dock.

Warping a crane into position to place a new section on strong backs off caisson seat.

11

Out with the old, in with the new

12

Realigning the vessel

13

Removing Scrap

14

Questions?

15

CUTTING THE

FLOATING DRYDOCK

MACHINIST

Presented at:

DM CONSULTING
DRY DOCK CONFERENCE 2017

Houston, Texas
May 5-6, 2017

Presented By:

Robert Heger
HEGER DRY DOCK, Inc.
531 Concord Street
Holliston, Massachusetts

TABLE OF CONTENTS

1. INTRODUCTION

The Floating Dry Dock "MACHINIST" was taken out of service in November, 2010 to undergo an extensive renovation program to extend its useful life and increase its rated lift capacity under MIL-STD 1625D.

During the renovation process the dock suffered a major casualty and sank at it's berth at Apra Harbor, Guam.

The dock was successfully salvaged and refloated a few months later.

Some time after the refloating it was decided to complete the renovations in China where construction costs are significantly lower.

Due to the deteriorated condition of the dock and damaged incurred during the sinking it was not possible to wet tow the dock from Guam to China. In addition, the dock was too long to fit on most heavy lift vessels. It was decided to cut the dock in half while afloat in its berth and load the two halves of the dock onto a single heavy lift vessel (with their longitudinal axis turned 90 degrees to the longitudinal axis of the vessel).

This paper discusses the procedures used for cutting the dock while afloat at its normal mooring berth.

2. OVERALL CONCEPT OF CUTTING THE DOCK

The dry dock was to be cut while afloat in its normal mooring berth (moored with a spread anchor chain system). The cutting was to be performed during a period of calm weather.

Ideally, the dock was to be cut into two approximately equal sections to minimize the length of the sections which would over hang the sides of the heavy lift vessel by quite a bit . Because the control house, electrical panels, stairways, and other equipment located around the dock's amidships (Frame 168) the cut was made at Frame 156, somewhat off centerline.

In order to insure no relative movement between the sections as the cut was made, the dock was ballasted in such a manner as to result in the draft, list and trim of each resulting section being the same before the cut as after the cut. The required ballasting was calculated by conducting a detailed weight survey of all weights and location of weights on the dry dock. Using this information a ballasting plan was developed to result in no calculated bending moment or shear force at the location of the cut. This insured the individual sections would theoretically float at the same draft, list and trim after the cut as they were prior to the cut i.e. no relative movements between the sections during the cut.

Because this was a theoretical calculation that depended on the accuracy of the weight survey, it was not realistic to assume it was perfect and no movement of the sections during the cut would occur. For this reason the cut was made in a precise order which would gradually reduce the longitudinal strength of the dock and allow any relative movement of the section to develop slowly, be accurately monitored and ballast adjusted to compensate. Relative movement between sections would indicate bending stresses still existed at the cut location and ballast was adjusted to eliminate this bending stress by bringing the dock back to its original precut deflection. A portable ballasting system was utilized to adjust ballast in certain tanks as the cut were being made.

If vertical shear existed at the cut line there would be no relative movement to indicate this until the load exceeded the capacity of the remaining plate and the plate failed. This could cause a vertical movement between sections and could be dangerous. For this reason shear beams, which extended across the cut line were added at the pontoon deck level. The shear beams would prevent vertical movement between the sections but allow the sections to be pulled apart after the cut is made.

The dock was moored by a spread chain mooring system with chains in the longitudinal and transverse direction. After the dock was cut in two, the tension in the longitudinal chains would tend to pull the sections apart. For this reason wire rope stops were added across the cut. The wire rope would allow the sections to pull apart a set distance but not too far to cause a problem.

3. PREPARATION OF THE DOCK PRIOR TO CUTTING

3.1 Weight Survey

A detailed weight survey was conducted on the dock prior to cutting. At the time of the cutting Guam Shipyard had a large amount of steel plate, steel members and equipment on the pontoon deck to be used in the repair of the dock. This weight had to be accounted for and added to the light weight of the dock.

3.2 Draft Measurements

The dock drafts were measured to within +/- ½". This was used to calculate the dock's displacement and was compared to the weight estimate. This insured the weight estimate and the actual dock's displacement were the same.

3.3 Deflections

Approximately 1 week before the cutting was to take place, Guam Shipyard began measuring the dock's longitudinal deflection along both wing walls at amidships, end and quarter points a few times per day.

At the time the deflection was taken the weather, temperature and daylight conditions were noted (cloudy, rainy, sunny, dark, etc.).

Any significant weight additions, removals or shifting that had occurred on the dock since the last deflection readings were recorded.

This provided a history of how the deflection changed throughout the day and with varying degrees of sun. This was used to determine whether or not changes in deflection during the cutting was due to the cutting or due to normal variations in temperature of the steel.

3.4 Shear Beams

Prior to cutting the dock, shear beams were installed across the cut line. On one side of the cut line, the beams were welded to the pontoon deck. On the other side the beam just rested on the deck. The welded side alternated on either side of the cut. In this way, the beams would prevent one side or the other from dropping down relative to the other after the cut was made but would still allow the sections to be pulled apart after the cut.

Shear Beams on Pontoon Deck

3.5 Cut Line Measuring Devices

It was important to measure the movement of the dock as the cut progressed. The intent was to determine whether or not the cut at the top of the wing wall and/or the cut at the pontoon deck was opening up or closing up as the cuts are being made and the dock was losing longitudinal strength. Any movement across the gaps indicated that the dock was deflecting and not ballasted properly. This meant ballasted needed to be adjusted. The measuring devices, called "Gap O Meters," were installed across the cut line at the following locations:

- Bottom of inner shell (just above pontoon deck), Port Wing
- Wing deck – Port wing
- Bottom of inner shell (just above pontoon deck), Starboard Wing
- Wing deck – Starboard wing

Gap O Meter

During the cutting, these measuring devices were monitored for movement. If movement was detected the cutting was stopped and the ballast adjusted to bring the movement back to zero.

3.6 Modification of Non-Watertight Bulkheads

Once the pontoon bottom plate was cut water would flood into the ballast compartments in which the cut was made. In order to minimize the extent of this flooding, the non-water tight bulkheads on either side of the cut were made water tight. This limited the flooding to one 5 meter long bay across the width of the dock.

3.7 Temporary Ballast System

The dry dock's existing ballast system was not operational. A temporary ballast system was installed to enable ballasting and de-ballasting the following tanks:

- Ballast Tank 1
- Ballast Tank 4
- Ballast Tank 13
- Ballast Tank 16
- Ballast Tank 21
- Ballast Tank 24
- Ballast Tank 37
- Ballast Tank 40

These 8 tanks were selected because they were the four corner tanks of the resulting sections after the dock was cut. Ballast in these tanks would have the greatest affect for adjusting list and trim in the resulting sections.

3.8 Develop An Initial Ballast Plan For Dry Dock

Heger Dry Dock developed an initial ballasting plan based on the weight survey and draft measurements. The plan put the dock in a zero stress condition at the cut line (theoretically). The dock was pre-ballasted to this condition prior to cutting.

During the cutting, the ballast would be adjusted to offset longitudinal deflection or gap movement observed during the operation.

4. CUTTING THE DOCK

4.1 Step 1 – Cutting Internal Longitudinal Members in the Pontoon

The first step in the cutting was to cut the internal longitudinal members in the pontoon. This consisted of the longitudinal bulkheads and longitudinal tee beams on the pontoon bottom plate and under the pontoon deck.

A 2 foot section of each tee member was cut out with the cut out length centered on the theoretical cut line. This enabled the underwater cut of the bottom plate, which would be performed by divers, to vary considerably from a straight line while still falling within the cut out section.

The longitudinal members needed to be cut out for a length greater than the accuracy of the diver's straight line cut. (i.e. The underwater cut line of the bottom plate must fall within the gap cut out of the longitudinal members above or the dock will not separate. If this happens, it would be hard to go back into the pontoon and recut the longitudinal members as that bay would be flooded.)

The divers said they could easily make the cut within the 2' accuracy.

A vertical cut in each longitudinal watertight bulkhead was made. At the top and bottom of the cut line an area 2 foot long, extending from the bottom plate up 6 inches was.

During the cutting, the dock's longitudinal deflection was measured at approximately once every hour. The deflection did not change.

4.2 Step 2 – Cutting the Underwater Shell Plate of the Pontoon

The initial plan called for the bay encompassing the cut line to be filled with water to the level of the outside water. This was to prevent water from rushing into the compartment when the initial underwater cut was made. However the divers requested the tank to be empty when starting the cut as they wanted the difference in pressure to suck the water into the tank and clear the area of the cut. They also used rubber seals on the cut to slow the flooding of water into the tank and keep this differential as long as possible. Portable pumps were also used to keep some differential between outside and inside water in the bay where the cut was being made.

The divers cleaned the area of the external shell plate that was to be cut and then clearly marked the line of the cut across the bottom and up the sides to a height of 1 foot above the waterline.

Starting from about 1 foot above the waterline on one side of the dock the divers cut the pontoon side shell, continued across the bottom plate and then up the other side shell to a height of about 1 foot above the waterline.

Start of Underwater Cutting

After the underwater portion of the plate was cut, the cut was independently checked visually and by running a knife blade in and along the entire length of the cut. This verified that everything had been cut.

During the cutting, the dock's longitudinal deflection was measured at approximately once every hour. No change in deflection was seen.

4.3 Step 3 – Cutting the Above Water Shell Plate of the Pontoon

The line of the cut across the pontoon deck and down the sides to meet up with the underwater cut was clearly marked.

From the outside, the burners cut the pontoon side plates first. This cut started where the divers had stopped and continued up to the pontoon deck level.

Next the pontoon deck plate was cut between the wing walls.

After the cuts were made, the cut were independently checked visually and by running a knife blade in and along the entire length of the cut. This verified that everything had been cut.

During the cutting, the dock's longitudinal deflection was be measured approximately once every hour. The deflection changed slightly and this was not attributed to temperature differences so ballast was adjusted to bring the deflection back to the initial reading.

4.4 Step 4 – Cutting Internal Longitudinal Members in the Wing Wall

A 1 foot long section was cut out of all longitudinal HP's in the wing wall. This included the HP's on the side shells, safety deck, mezzanine deck and wing deck. The cut out length was centered on the cut line.

Next, the safety decks and mezzanine decks were cut. At the inner and outer shells, an area 1 foot long, extending from the shell plate in 6 inches was cut in the deck plate.

All pipes, wires and other items that spanned across the cut line were disconnected at the cut line.

During the cutting, the dock's longitudinal deflection was be measured approximately once every hour. The deflection changed slightly and this was not attributed to temperature differences so ballast was adjusted to bring the deflection back to the initial reading.

4.5 Step 5 – Cutting the Wing Decks

The line of the cut across the wing deck was clearly marked.

Prior to cutting it was verified all pipes, wires and other items that spanned across the cut line had been disconnected at the cut line.

The crane rail was cut out in the area of the cut line.

From the outside, the burners cut both wing decks simultaneously.

During the cutting, the dock's longitudinal deflection and gap measuring devices were monitored continuously. The gap on the pontoon deck had opened up slightly and the deflection had changed slightly. This was not attributed to temperature differences so ballast was adjusted to bring the deflection and gap measurement back to the initial reading.

4.6 Step 6 – Final Cut - Cutting the Wing Side Shells

At this point the only structure joining the two sections was the vertical side shell plates.

The line of the cut down the wing side shells was clearly marked.

All four wing shells were cut simultaneously. From the outside the burners started cutting the wing shells in the following order:

1. From the pontoon deck up 5 feet, inner and outer shells, P & S
2. From the wing deck down 5 feet, inner and outer shells, P & S
3. From 5 feet to 15 feet above pontoon, inner and outer shells, P & S
4. From 5 feet to 15 feet below wing deck, inner and outer shells, P & S
5. From 15 feet to 25 feet above pontoon, inner and outer shells, P & S
6. From 15 feet to 25 feet below wing deck, inner and outer shells, P & S
7. Final cut, inner and outer shells, P & S

This pattern was selected to gradually reduce the longitudinal strength of the dock. If any bending stress was still in the plates at the cut this would show up as increased deflection (and movement across the gaps at the pontoon and wing deck levels). This could be monitored and corrected by adjusting ballast. At the final cut there should be no stress thus no ripping or tearing of plates.

An air horn signaled the start of cutting the next section. All four burners continued at a steady pace until the 5 foot increment had been cut. Then they stopped and waited for the signal to start the next section.

Final Cutting of Wing Shells

During the cutting, the dock's longitudinal deflection and gap measuring devices were monitored continuously. After about half the cutting was complete, the gap on the pontoon deck had opened up slightly and the gap on the wing deck had closed slightly and the deflection had changed slightly. This was not attributed to temperature differences so cutting was stopped and the ballast was adjusted to bring the deflection and gap measurement back to the initial reading. After resetting the deflections and gap measurements the cutting was resumed.

After the final cut was made nothing happened and for a few seconds it was thought we may have missed a cut somewhere and the dock was still connected. Then the sections started drifting slowly apart until the stop cables tensioned up.

Dock Gap After the Cut

At the final cut there was no popping or tearing of plate nor any significant movement between section. This indicated the sections were perfectly balanced at the time of the final cut.

HEGER DRY DOCK, Inc.
May 2017

Shipyard Solutions

TTS Syncrolift

Introduction

TTS Syncrolift has during the last thirty years been a solutions provider to shipyards worldwide. Our solutions are characterized by the focus on productivity and logistics.

We are today the world's leading supplier of Shiplift and transfer systems. Our clients are:

- **Navy yards**
- **Commercial repair yards**
- **Newbuilding yards.**

Our vision for the future - is to continuously introduce new solutions which will improve productivity and lead time for the repair of ships.

TTS Syncrolift has three main offerings to our clients:

- **Docking solutions, Transfer Systems**
- **Fast Docking Tools** (tools which improve speed of repair)
- **Services for existing systems in operation**

TTS Syncrolift Products overview

TTS Syncrolift - installed base worldwide

TTS Syncrolift AS - Support worldwide

Norway, Drøbak
TTS Syncrolift AS
Main office – all queries
T +47 64 90 79 10
E service.syncrolift@ttsgroup.com

USA, Arnold, MD
TTS Marine Inc
Dept office – all queries
T +1 410 093 0403
E service.syncrolift@ttsgroup.com

Singapore
TTS Singapore Pte Ltd
Service office - After Sales & Services
T +65 6867 9070
E service.syncrolift@ttsgroup.com

Vietnam
TTS Vietnam
Service office – Services & support
T +84 913 538 800
E service.syncrolift@ttsgroup.com

Emergency technical support, 24 hours
T +47 64 90 79 20

2

3

Docking Solutions

Overview Docking solutions products

Every location has its advantages and disadvantages in terms of soil, water debt and so on. Being a solutions provider, we must therefore offer different solutions which better fit the local conditions.

TTS Syncrolift offer the following solutions:

- **Syncrolift® (shiplift)**
- **Slipway – wedged slipway**
- **Floating Dock**
- **Docklift**

Shiplift Port Adelaide, Australia

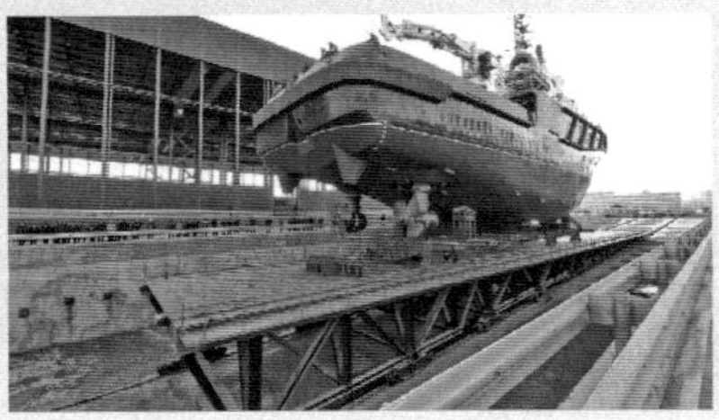

Slipway, Polo Russia

Control system and unit

All of our docking solutions have a control system with which to operate the equipment. The Syncrolift® is a good example of how TTS Syncrolift optimises Human Machine Interaction and incorporates necessary equipment in a thoroughly designed user experience, both when it comes to the visual and user friendly aspects.

The Syncrolift® control tower is usually located at the shore end of the shiplift and elevated above ground to give a good overview of the docking procedures.

The control desk is specially designed to be easy for the user to interact with, and at the same time have one eye on the operation, both on cameras and out the windows.

In addition to lifting/lowering platform, some features of our Control system are:

- **Distribute shipload on platform** (allows winches to share loads and collectively lift higher loads)
- **Safely Locking platform**
- **Alarms systems ensuring fail safe operation**
- **Tilting and listing of platform in case of ships with uneven keel or uneven ballasting**

Our control tower with Control system is designed for a "plug and play" installation.

(For more information, see Docking Solutions brochure.)

An artist's rendering of a control tower on order by one of our customers

Moving a ship in to a floating dock

4

5

Ship Transfer Systems

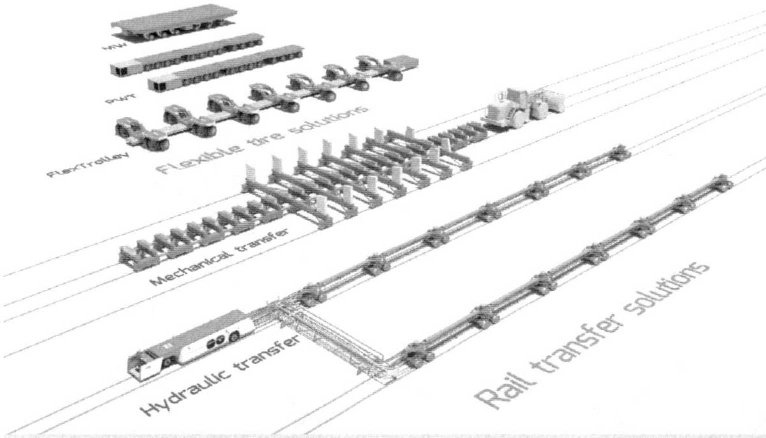

Overview transfer systems products

TTS Syncrolift has designed and delivered transfer systems world-wide for the last thirty years. This experience is accumulated into our transfer systems.

A challenge for most yards is the peak-load characteristic of many ships. i.e. a tendency where parts of the ships have a very high load. The problem this creates is higher investment in ground or shiplift. Reason being that peakload becomes the design criteria.

TTS Syncrolift has a solution to this Our fluid bed technology is designed in such a way to share loads between each trolley. This way - a peak load will be averaged down and hence we lower peak-load and therefore investment.

The following is a list over our different transfer systems:

- **Fluid-bed Transfer system** (rails – self-propelled)
- **Rigid Transfer system** (rails)
- **FlexTrolley** (Tires - self-propelled)
- **Power Wheel Transporter PWT** (Tires-self-propelled)
- **Multiwheeler MW** (Tires - self-propelled)

Most of these systems require that the ship/equipment is supported on some type of trestle, making it possible for the transfer system to get under the ship to lift it up. You can see 3 types of a trestles in the pictures to the right. It is the construction between the ship and the transfer system.

Tire, rails and Layouts

There are two main groups of transfer systems. These are:

- **Rail transfer solutions**
- **Flexible tire solutions**

The two pictures below show the same shipyard with rail transfer solutions and flexible tire solutions. A yard layout should be designed to maximize the number of vessels it can dry dock, while keeping its flexibility to dock/undock vessels.

Tire based transfer solutions can in many cases allow more freedom to maximize the space-capacity of the yard, because rail transfer solutions move mainly in x and y direction.

Tire-based solutions can move in any direction, including rotation. This means it may also get past buildings or into halls that are a little off the path. With tire-based transfer solutions, it is possible to dock more vessels in your yard than with a rail-transfer solution.

Rail-transfer solutions have a greater lifting capacity, so they can transfer heavier vessels than tire transfer solutions.

TTS Syncrolift can help you find the best solutions for your yard.

(For more information, see Transfer Systems brochure.)

Layout Rail transfer solutions

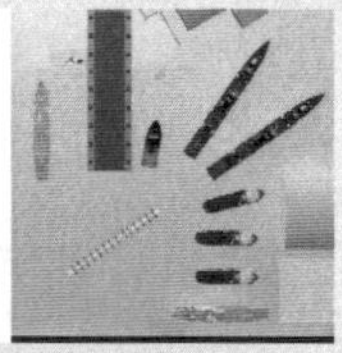

Layout flexible tires transfer solutions

PWT. Dammen Song, Vietnam

FlexTrolley

Hydraulic transfer system

Reduced Docking time

Overview fast docking products

Ship owners make their money from continuously sailing cargo. A repair yard with low lead time and competitive prices are attractive to most ship owners. Most repair yards today are based on manual work. This requires a relatively high number of workers and it requires more lead-time for repair.

This is a challenge for repair yards. For TTS Syncrolift it is an opportunity to find solutions which reduces man-hour cost and lead time. This business area is called Fast Docking Products.

Typical operations where manual work is still the standard within repair yards are:

- **Preparation work for ships such as blocking for keel/bilge support**
- **Propeller repair**
- **Manually inhauling ships for docking**

TTS "Fast Docking Products" are:

- **Multi Vessel Docking Systems**
- **Bilge Support Arms**
- **Inhaul & Positioning Systems**
- **Dock Side Support Arms**
- **Dock Arms**
- **Auto Block Guiding**
- **Propeller Puller**
- **Flex Pad**
- **Shiplift Side Support Arm**

Innovative solutions

Streamlining docking time is crucial for the owner of the vessel and gives the yard a good reputation. The only limits to improving working processes or finding new solutions in this product group, are one's knowledge and imagination.

TTS Syncrolift is moving this product group forward by continuously focusing on cooperating with our customers. In this way we accumulate the knowledge from the users to invent/develop new products and improve existing ones. The best way to maxmize innovation is to accumulate our know-how with our customers' experience. This process will make the products more intuitive, faster and safer to use, and will reduce the docking time for a vessel.

Our vision is to reduce docking times significantly through our fast docking products, and make the products safer and more intuitive to use.

(For more information, see Fast Docking Systems brochure.)

An artist's rendering of a inhaul and positioning system, to one of our costumers

An artist's rendering of a propeller puller to one of our costumers

Dock Side Support Arms, [illegible], Norway

After Sales & Services

TTS Syncrolift AS, the OEM supplier of Syncrolift® shiplifts can provide you a full suite of services. Service range from long-term support solutions, to modular software upgrades, capacity upgrades, to spares provisioning, spares inventory, periodic inspections, maintenance and repairs etc.

With technical support and skilled engineers based around the world, support is available when and where you need it.

Upgrades & refurbishments

Upgrading older systems can significantly improve performance and safety. The latest control systems provides up-to- date functionality regardless of age. This control system is intuitive and easy to understand and learn. As technology has become more intricate and users want more features, the task is keeping things simple.

Global distribution - OEM spares parts - Maximizing safety

Genuine OEM spares is supported by a global distribution network to provide the assurance, quality and availability your operations demand. Rigorous design and quality standards are applied to all safety-critical components. TTS Syncrolift strongly recommends employing only original OEM spare parts and wire ropes on your Syncrolift Shiplift and transfer system installations.

Safety- critical spares as wire ropes are designed by TTS Syncrolift engineers to suit the unique characteristics of the Syncrolift® shiplift system. Manufactured by one carefully selected wire rope manufacturer to TTS Syncrolift exacting specifications. Syncrolift® wire ropes and OEM spares ensure that the Syncrolift® shiplifts perform to the characteristics intended, while maximizing the safety of the system. Further, there are over forty Syncrolift® shiplift installations that are owned and operated by government organizations and navies throughout the world, including two facilities used to launch/dock nuclear submarines. These installations follow strict procurement guidelines and procedures. This leads those to strictly procure OEM spares and Syncrolift® wire ropes as a sole- source item. They cannot afford to take a chance on a substandard product.

10

Through life support, training and inspections

Our shiplifts vary in size and build (articulated and / or rigid) which may affect your needs. There are various levels of through- life support that can be tailored to each Syncrolift Shiplift installation, allowing customers to choose the programme that suits them best. Regular inspections and training are key in maintaining the operational safety of the Syncrolift® shiplift and transfer system. This will maximize availability and reduce overall maintenance costs.

TTS Syncrolift offers comprehensive inspections where our service engineers reviews previous maintenance, repairs, and modifications since last inspection. During an inspection, we offer a large scope of services. One important task in this scope is NON-destructive examination of the Syncrolift® wire ropes. The inspection will then results in a final detailed report of the Syncrolift® shiplift installation condition including recommendations for repairs and improvements.

We will work with owner/ management to provide you the best support.

TTS Syncrolift AS has safety as or top priority. We provide comprehensive operator, dock-master and maintenance training and a range of courses are available that can be delivered on-site to meet your specific needs.

Simulator training

In addition to our traditional training of new and experienced users of TTS Syncrolift equipment, we offer training by use of our simulator, which will give a more realistic understanding of the equipment/ systems for a user/dock- master. Simulation training programs are available for Syncrolift® (shiplifts), transfer systems and maintenance of the equipment. The training is suitable for all personnel in contact with TTS Syncrolift equipment.

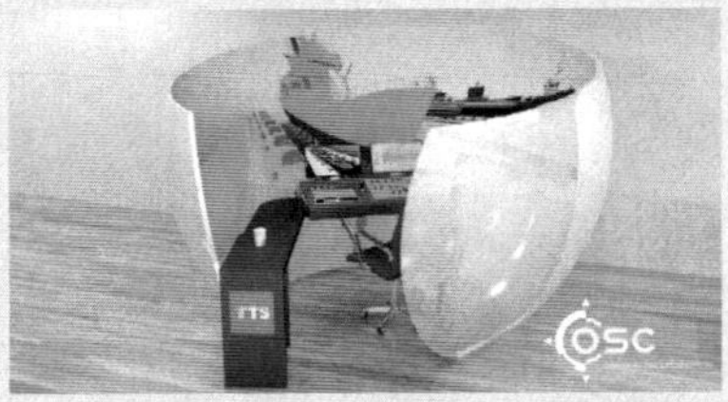

Simulation training

Technical support

We understand the value of speaking to someone to help manage your day-to day issues. Each Syncrolift shiplift installation and transfer system has a dedicated project manager on hand to answer any technical enquiries and manage work requests.
TTS Syncrolift offers remote technical support on demand. We can connect through a PLC communication module and/or directly to the Syncrolift® shiplift PC. By remotely connecting to your system, we can provide system diagnostics, remote fault-finding, system monitoring, and operator support. We also offer to upgrade existing systems with remote features if this is not already enabled/available on your system.

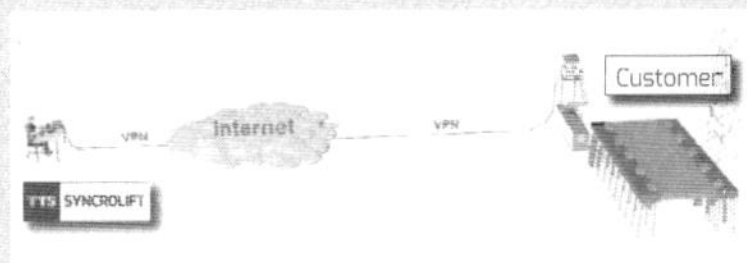

After Sales and Services contact
T Office: +47 64 90 79 10
E service.syncrolift@ttsgroup.com

Emergency Service contact (24 h)
T +47 64 90 79 20
(See page 3 for worldwide contacts)

11

SMC
SCHULTE MARINE CONCEPT

Floating Dry Dock newbuilding project in China

The 9th Dry Dock Conference

5-6 May 2017, Houston, TX, USA

shipbuilding sector in China

Background … 19th century

- accumulated experience
- technical culture
- associated services and manufacturing industry

- over 1,400 harbors, over 210,000 cargo ships

SMC

shipbuilding sector in China

Capacity

- 65m dwt 2015 (2012: 80m dwt)
- > 3,000 shipyards 2010
- 71 shipyards ‘White Listed’, September 2014
- state owned yards / CSSC, CSIC
- join venture foreign and domestic yards
- private owned yards / stock listed

shipbuilding sector in China

Capability

- all type of merchant vessels
- military units submarines, surface combatants, naval aviation, sealift
- all type of <u>offshore units</u>

shipbuilding sector - ??????

- one yard
- price breakdown

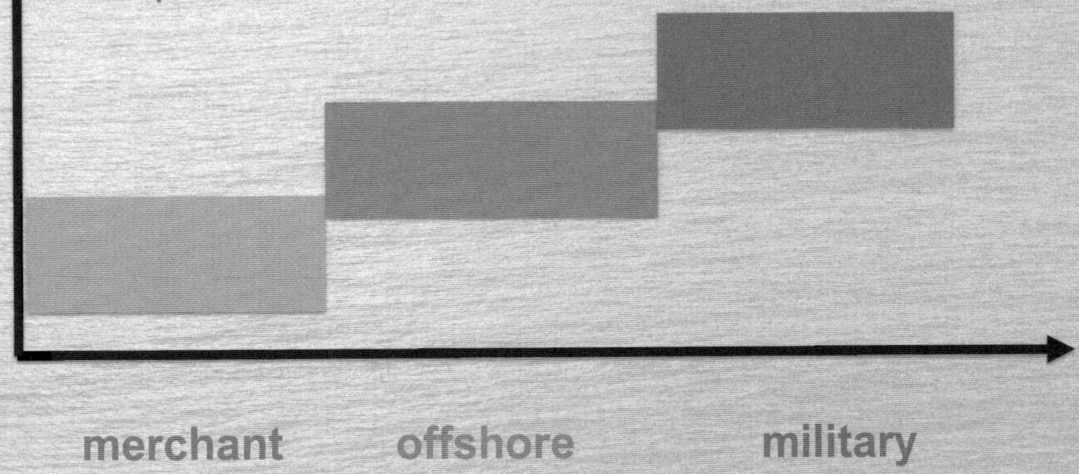

SMC

SCHULTE MARINE CONCEPT

shipbuilding sector in China

Shipyard's clientele:

- tonnage providers / shipowners
- commodity majors & traders
- cargo forwarders
- investors
- … somebody needs FDD

SMC

SCHULTE MARINE CONCEPT

FDD newbuilding project - design

Design … *YCGWYWIYKWYW* you get what you want …

- own design

- shipyard's design

SMC

Schulte Marine Concept

FDD newbuilding project - design

Design … *YCGWYWIYKWYW*

- own design developed by Buyer, customized by Builder
  - ✓ intrinsic technical & operational compliance
  - ✓ basic → Class → … woz straz → detail → 3D → production + MTO
  - ✓ 3D model review

SMC

Schulte Marine Concept

FDD newbuilding project - design

Design ... *YCGWYWIYKWYW*

- own design developed by Buyer, customized by Builder
  - ✓ intrinsic technical & operational compliance
  - ✓ basic → Class → ...woz straz → detail → 3D → production + MTO
  - ✓ 3D model review
- shipyard's design developed by Builder, customized by Buyer
  - ✓ technical & operational compliance challenge
  - ✓ basic → Class → detail → 3D → production + MTO
  - ✓ plan approval and 3D model review

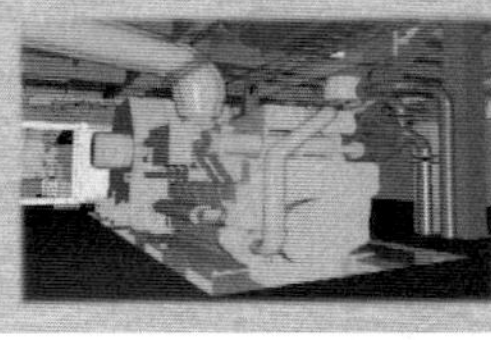

SMC

FDD newbuilding project – makers list

Makers List shortest , but most expensive document

- Building Contract
- CAPEX & OPEX
- trading pattern
- shipyard ... domestic production
- tailor design, but standard equipment
- it is all about details ...Technical Agreements

SMC

FDD newbuilding project - shipyard

Shipyard … multicriteria selection process

- state owned / jv/ private
- price
- track record
- infrastructure
- geographical location
- $ & orderbook
- gap analyses / FMEA

FDD newbuilding project – shipbuilding contract

Shipbuilding contract … *YCGWYWIYKWYW*

- Ship Building Contract
- Technical Specification
- Makers List
- scene setting

shipbuilding sector - ??????

- one yard
- price breakdown

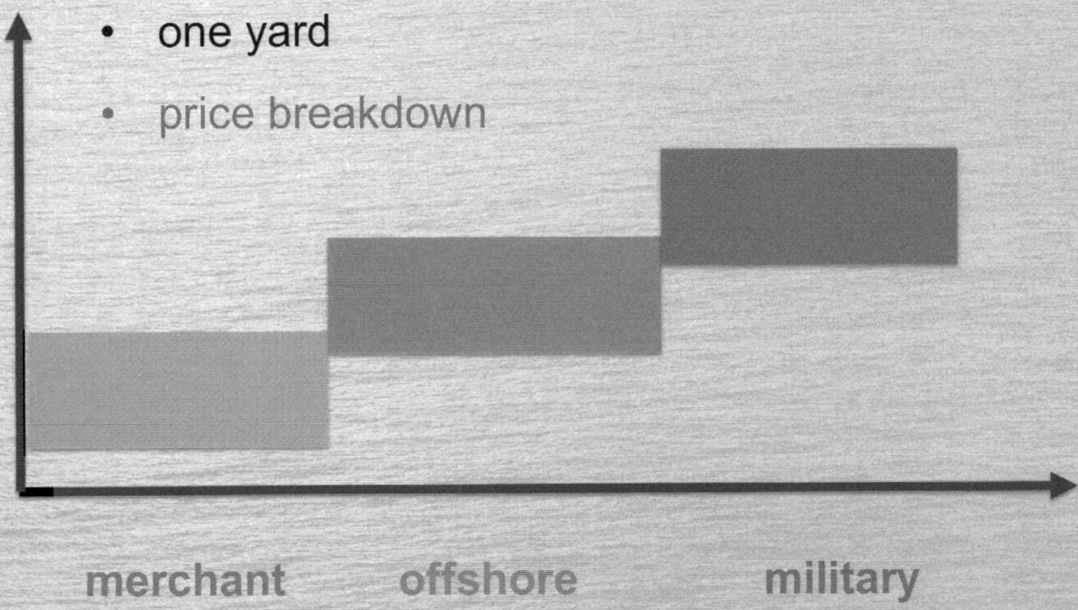

SMC

FDD newbuilding project – site supervision

Supervision ... *YGWYI* you get what you inspect

- HSE
- compliance: preventive & corrective
- subcontractors' audit
- equipment storage / conditions & maintenance
- cleanliness & protection
- troubleshooting v. accountability

(turn key SBC)

SMC

FDD newbuilding project – towage

Towage

- turn key contract

SMC

Schulte Marine Concept

FDD newbuilding project – shipbuilding contract

Final assembly & commissioning

- SBC
- design & production
- time
- Builder's technical support

SMC

Schulte Marine Concept

FDD newbuilding project

GLOBAL KNOWLEDGE – LOCAL SOLUTIONS

SMC

SCHULTE MARINE CONCEPT

people are different

SMC

SCHULTE MARINE CONCEPT

Schulte Marine Concept

Thank you

web: www.schultemarineconcept.com

SMC
Schulte Marine Concept

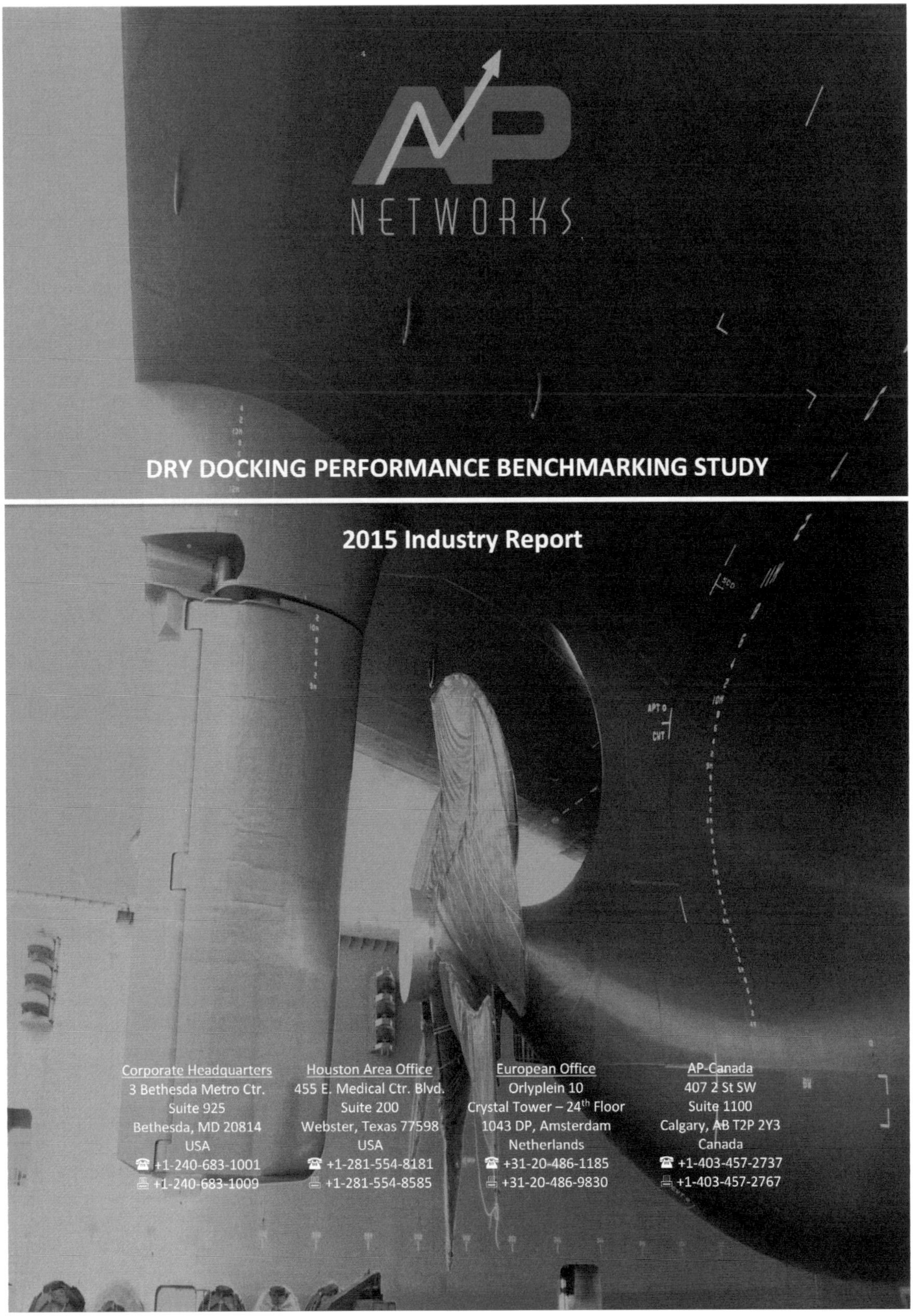
AP
NETWORKS
DRY DOCKING PERFORMANCE BENCHMARKING STUDY
2015 Industry Report
Corporate Headquarters
3 Bethesda Metro Ctr.
Suite 925
Bethesda, MD 20814
USA
☎ +1-240-683-1001
🖷 +1-240-683-1009
Houston Area Office
455 E. Medical Ctr. Blvd.
Suite 200
Webster, Texas 77598
USA
☎ +1-281-554-8181
🖷 +1-281-554-8585
European Office
Orlyplein 10
Crystal Tower – 24th Floor
1043 DP, Amsterdam
Netherlands
☎ +31-20-486-1185
🖷 +31-20-486-9830
AP-Canada
407 2 St SW
Suite 1100
Calgary, AB T2P 2Y3
Canada
☎ +1-403-457-2737
🖷 +1-403-457-2767

Summary

The 2015 Dry Docking Performance Benchmarking Study was conducted to provide participating companies with:

- Direction on where and how to improve their dry docking performance.
- Quantifiable information to support the decision-making process regarding when and where to execute dry dockings.
- Facts related to location, costs, schedule, and quality to confirm or refute current strategies.
- Dry docking performance benchmarks that can be used in discussions with ship owners and commercial partners.
- Actionable recommendations regarding how to capitalize on opportunities and improve dry docking performance.

In the preceding months, all participating companies were briefed on their individual dry docking performance. Observations and recommendations based on the data collected and analyzed, and on the information regarding dry docking practices received during face-to-face interviews, have been shared.

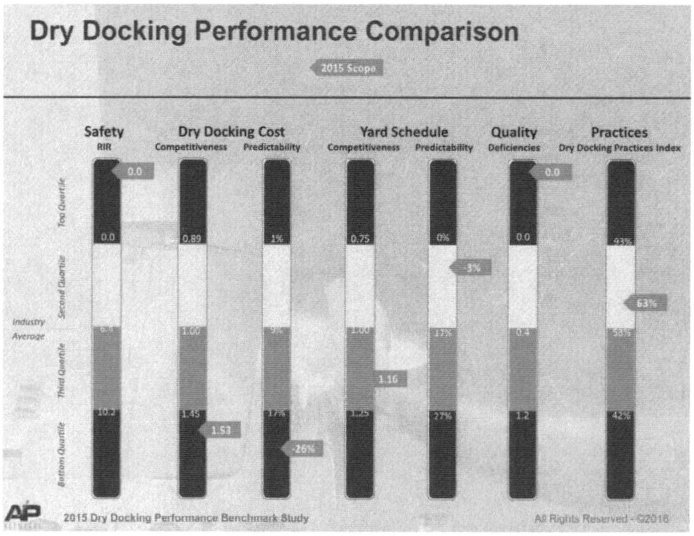

Figure 1: 2015 Dry Docking Performance Benchmarks

Study Background

The Dry Docking Performance Benchmarking Study was conducted by AP-Networks, Industry's premier turnaround benchmarking and assessment firm.

AP-Networks facilitated the study, ensured alignment with Industry standard definitions, conducted analysis and interviews, and served as the custodian of study data.

Collected dry docking data has been validated, evaluated, and benchmarked against the AP-Networks Dry Docking and Upstream Turnaround Database, the most comprehensive database of dry docking observations in the world.

Participating study sponsors included a diverse array of ship management and owner organizations.

As shown in Figure 1, the 2015 study participants achieved top quartile performance results for both Safety and Quality metrics. However, the Safety score exhibits a significant caveat, as it represents only the results of the dry docking teams. It is remarkable in the modern era of health and safety focus, that the owner/ship manager organizations do not systematically track overall shipyard workforce health and safety performance.

Furthermore, overall Cost Competitiveness for study participants is bottom quartile, with an average cost competitiveness index of 1.53 (see explanation of index in sidebar). In addition, Cost Predictability was minus 26 percent, also bottom quartile. As this is measured on actual performance data, this result indicates inadequate cost estimation and target setting.

Schedule performance showed a slightly better outcome, with a third quartile schedule competitiveness index of 1.16. This result can be considered acceptable, especially in light of the current economic environment, though not exemplary. In such a market, predictability is key to success. Study participants achieved Schedule Predictability of three percent, which is second quartile performance.

Competitivenessare and Predictability are directly correlated with the quality of the planning and preparation phase. Following the review of planning deliverables and subsequent interviews, a Dry Docking Practices Index was calculated. The 2015 study participants achieved a Practices Index of 63 percent. Although slightly better than Industry average, this is still significantly less than an optimal score of 93 percent.

Based on the performance analysis, the following recommendations can be made:

- Track yard safety statistics.
- Start the planning and preparation phase of dry dockings earlier to avoid expensive decision making amid pressure from the yard contractor.
    - During the planning and preparation phase, place greater emphasis on cost and schedule execution competitiveness.
- Improve the scope collection process:
    - Formalize scope criteria, as well as scope addition and deletion processes.
    - Formalize the Risk Based Scope Review process.
    - Clearly mark opportunity scope and define the conditions required for scope to be included during dry docking execution.

Key Terms

Competitiveness is a measure of performance relative to the Industry dataset average.

$$\frac{\text{Actual Cost and Duration}}{\text{Industry Average Cost and Duration}} = 1.0$$

The Industry average and top quartile numbers are generally derived from the AP-Networks database and generated using a statistical model that normalizes for unit type, complexity, and location.

In those cases where the **Industry Average** is not generated from an AP-Networks statistical model, averages of the comparative datasets were used.

Predictability is defined as the difference (expressed as a percentage) between the total actual cost or schedule and the planned cost or schedule estimate.

$$\left(\frac{\text{Actual Cost and Duration}}{\text{Planned Cost and Duration}} - 1\right)$$

- Actively control the scheduling process by requiring an approved schedule before the start of execution.
- Review the hull repair scope identification process with the objective of achieving a more accurate and less conservative estimate of hull repair scope.
- Increase the involvement of a procurement specialist in the yard selection process in order to build greater procurement knowledge and to free up time for the Technical Superintendents.

Fast Yard Schedule versus Low Costs

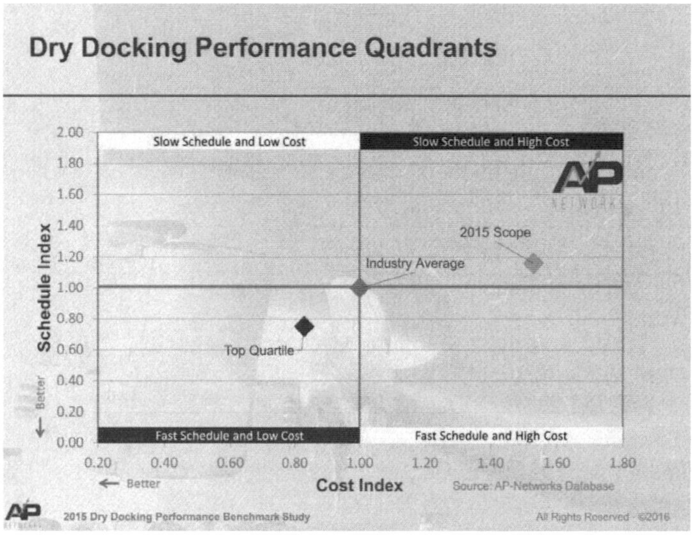

Figure 2: 2015 Dry Docking Performance Quadrants

Looking at the dry docking performance quadrants highlighted in Figure 2, it's clear that the bottom left quadrant illustrates an ideal scenario (i.e., Fast Yard Schedule and Low Cost). However, 2015 study performance landed squarely in the top right quadrant (i.e., Slow Schedule and High Cost), showing that there is still a long way to go before the study participants achieve optimal dry docking performance.

At AP-Networks, we have watched organizations in a wide array of industries move across this chart, and have firsthand knowledge of what it takes to make those moves. Our work in these industries has taught us that organizations can achieve vertical movement (i.e., Improving the Schedule Index) by spending money (e.g., on yard contractors). To achieve horizontal movement (i.e., Improving the Cost Index), planning and preparation need to improve (e.g., score on the Practices Index).

Definitions

Off-Hire Period: The time that a vessel is not available for hire; normally the period between the last port call before docking and the first port call after docking.

Yard Period: The time between arrival at the shipyard (berthed at quayside) and departure from the shipyard.

Docking Period: The period between the start of work by the shipyard and the completion of said work.

Contingency Allowances

As with many assets that are operated on a continuous basis, planned dry docking scope is not defined with 100 percent accuracy. To ensure that scope change does not immediately require additional budget requests, cost budgets routinely include contingency to a commensurate level of planning.

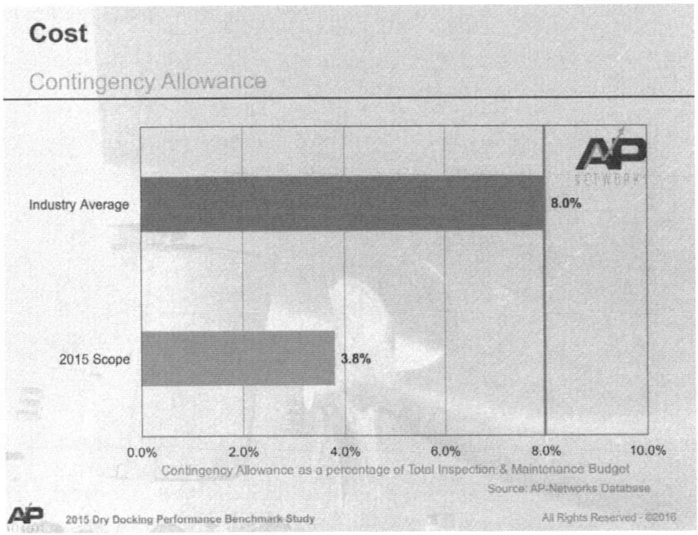

Figure 3: Cost Contingency Allowance

An average contingency of 3.8 percent was included in the dry docking budgets of the 2015 study participants, which is only half of the Industry average (i.e., 8 percent), as shown in Figure 3.

When looking at the current practices that define Cost Contingency, it is noteworthy that for the most part, contingency was only defined by the study participants based on historical figures and/or experience. At best, the age of the ship and its current reliability were also taken into consideration.

In-depth cost studies conducted by AP-Networks [1] show that organizations that define, manage, and monitor three different cost contingency allowances (i.e., emerging work, discovery work, and estimation errors) exhibit better cost predictability.

Definitions - Continued

Inspection Scope: Equipment inspection activities undertaken to assess equipment condition. This may include cleaning activities required to execute inspection, but it does not include any repair work.

Maintenance Scope: Activities related to equipment maintenance (e.g., breakdown repairs, preventive maintenance activities). This also includes any repair work resulting from inspection activities.

Ship Alteration Scope: Activities related to vessel upgrades that will influence ship characteristics (e.g., DWT, cargo holding capacity).

Ship Modernization Scope: Activities related to vessel upgrades that will not influence ship characteristics (e.g., anti-piracy measures, installation of emissions reduction systems).

[1] See http://www.ap-networks.com/download-document/28-GRL-Article-Cost%20Engineering.pdf

Scope Predictability

Defining the right scope for an upcoming dry docking is one of the most important Critical Success Factors, and should therefore be governed by Industry Best Practices, solid procedures, and firm approval processes.

With class societies, flag states, and OEMs providing clear guidelines, a significant part of dry docking scope seems predetermined. Nevertheless, there is a substantial portion that is condition based (e.g., condition of the hull painting, thickness of anchor chain).

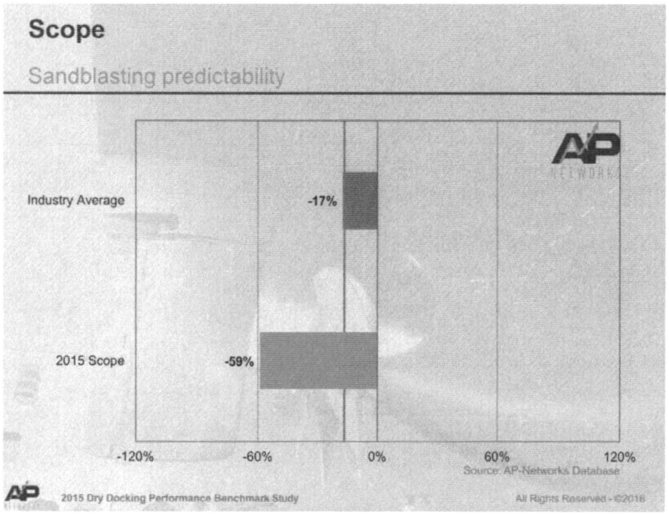

Figure 4: Scope Sandblasting Predictability

Normally, hull work starts after a full general hull inspection, and is executed by first sandblasting those areas that require work. As shown in Figure 4, Industry is conservative when estimating the required amount of sandblasting (-17 percent). However, the 2015 study participants were even more conservative (-59 percent).

In those cases where the hull work sits on the critical path, conservative estimation of the hull scope leads directly to a conservative schedule. AP-Networks' experience indicates that it is extremely difficult to fully realize schedule benefits resulting from scope reductions. Once a schedule is set, it is not typically underrun, regardless of a significant scope descrease. A better tactic for achieving competitive schedule performance, is to set a more competitive schedule from the beginning.

Definitions – Continued

Discovery Scope: Scope identified after the start of dry docking.

Emerging Scope: Scope identified in the period between scope freeze and the start of dry docking.

Baseline Budget: The total planned costs for the dry docking, excluding any contingencies and ship alterations or modifications.

Dry Docking Team: The team working on planning, preparing, and executing the dry docking. A team can consist of one or many individuals.

Dry Docking Work Process

In every dry docking work process, there exist Key Deliverables. These often require interaction with and approval from the ship owner, and serve as important milestones.

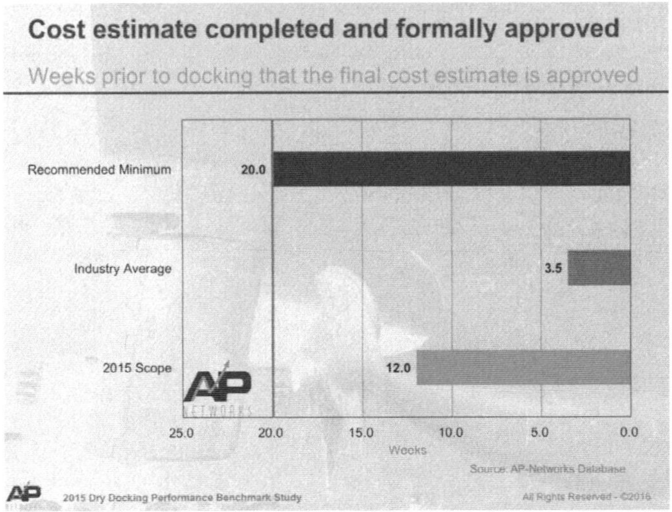

Figure 5: Cost Estimate Formally Approved

Formal approval of the cost estimate is a particularly important milestone in the planning and preparation process, as it allocates the budget and provides direction for the event. As shown in Figure 5, the Industry average of 3.5 weeks prior to docking reflects a very late approval milestone. At 12 weeks prior to docking, the 2015 study participants achieved significantly better performance on this metric. However, this is still much shorter than the recommended minimum of 20 weeks.

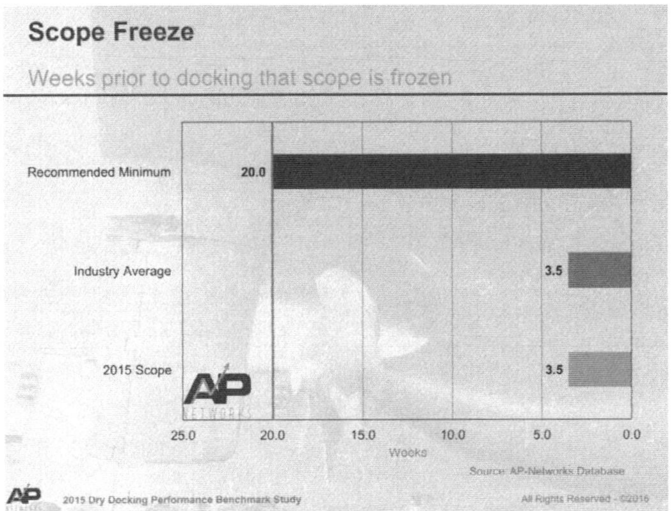

Figure 6: Scope Freeze

Benchmarking Methodology

Benchmarking requires a normalization methodology to make "like-for-like" comparisons. AP-Networks applies a two-step approach: First, the cost and schedule data are normalized to an equivalent basis to account for differences in region and complexity.

The second step is the application of a suite of cost and schedule benchmark models that are based on the outcomes in the AP-Networks database, and use Ordinary Least Square regression methodology:

$$Y = aX_1 + bX_2 + \ldots + \text{Constant}$$

These models allow us to identify and control for the key cost and schedule drivers so that comparisons across different types of dry dockings are made on an equivalent basis.

Freezing the scope is another key milestone, as it locks in the scope of work that should be executed. After scope freeze, only scope resulting from breakdowns and/or discoveries should be added. Figure 6 shows that Industry, as well as the 2015 study participants, freeze scope very close to the start of dry docking (i.e., 3.5 weeks prior).

One would expect that in order to complete a budget, scope would need to be frozen. As such, it is remarkable to see a mismatch between the data shown in Figure 5 and Figure 6. It raises the question, on which scope the budget is based and approved upon. This may offer an explanation for the significant cost underrun of -26 percent observed on the Cost Predictability metric.

Dry Docking Practices Index: 63 percent

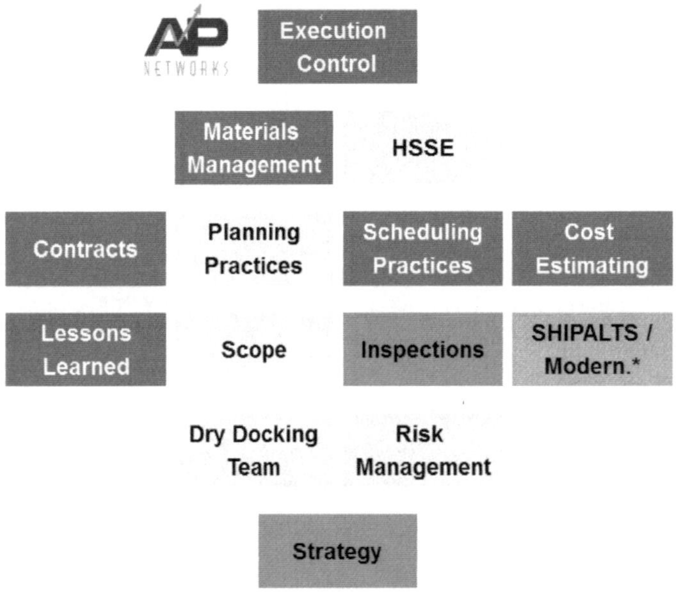

Source: AP-Networks Database

Figure 7: Dry Docking Practices Index

The AP-Networks database displays a relationship between the application of Industry Best Practices and dry docking performance outcomes. Improving the score on the Dry Docking Practices Index translates into improved competitiveness and predictability. Those improvements should be institutionalized through a dry docking work process, where Contracts, Lessons Learned, Materials Management, Scheduling, and especially Execution Control should be given top priority.

An Explanation of Performance Drivers

Performance Drivers are those elements that contribute to outstanding performance.

These performance drivers were obtained through a Practices Survey covering 14 different dry docking topics. This survey was completed by each of the dry docking teams in the 2015 study.

Some questions asked for a qualitative response using pre-defined answer options.

Other questions asked for quantifiable information, e.g. "How many weeks before the planned Dry Docking Start Date has the scope been frozen and a scope control procedure put in place?"

The best practice for each question was determined, receiving the maximum score. When combined with the survey score results, this information yielded a Dry Docking Practices Index score.

Some questions (e.g., Dominant Trading Area), have no best practice and are therefore excluded from the score calculations.

Information gathered during the interviews has been used to validate the answers given in the Practices Survey.

AP-Networks Dry Docking and Upstream Turnaround Database

With this study, AP-Networks is bringing its experience and proven benchmarking and data collection methodology to the maritime industry. This methodology has delivered powerful results in the offshore, petroleum, chemical, and energy industries, and to date has been employed successfully to aid various maritime companies including Rickmers, Masterbulk, and Transocean in improving their dry docking processes and performance.

Measuring, baselining, and tracking changes in dry docking performance is essential for long-term operational success. The AP-Networks Turnaround Database, with more than 1,350 data points for dry-docking-like events from various industries, contains information on planned and actual outcomes and key characteristics, as well as the planning and execution practices correlated with said outcomes. AP-Networks' Dry Docking and Upstream Turnaround Database is a subset of the Turnaround Database containing information from dry dockings.

Data analysis by AP-Networks facilitates insights into the causal factors behind success and failure, and allows individual dry dockings to be benchmarked against Industry average and top quartile performance. Our analysis also allows for the identification of demonstrated best practices that lead to competitive outcomes.

AP-Networks and the Maritime Industry

AP-Networks engages with dry docking and turnaround teams from offshore and shipping clients such as BP, Chevron, Neste, Masterbulk, Shell, Statoil, Suncor, and Transocean in order to help them achieve their goals in executing turnaround-like events.

The database holds information on ship types such as:

- Dry Bulk Cargo Vessels
- Container Vessels
- FPSOs
- Oil Platforms
- Drilling Rigs

Wage Rate, Regional Indices, and LPO

The AP-Networks database used for dry dockings is baselined as follows:

- US Dollars Jan 2016
- Dry dockings executed in the vicinity of Shanghai, China

Non-Chinese-vicinity dry docking labor components are normalized to Chinese wage rate equivalents using the information provided in the "UBS Prices and Wage Rates Indices Report 2015."

Lost Production Opportunity (LPO) cost has been used in a number of metrics not only to display costs directly related to a dry docking, but also to display lost revenue.

Vessel Type	LPO per Day
Handysize	US$ 4,000
Supramax	US$ 5,000
Container – 4,231 TEU	US$ 6,272

More information on the "UBS Prices and Wage Rates Indices Report 2015" can be found at: https://www.ubs.com/microsites/prices-earnings/prices-earnings.html.

About AP-Networks

Asset Performance Networks (AP-Networks) is Industry's premier turnaround benchmarking and assessment firm. We aim to advance the state of Industry knowledge, and to drive continuous improvement.

For more than a decade, our company has worked to understand and improve turnaround performance in the offshore, petroleum, chemical, and energy industries. By combining our detailed turnaround event knowledge with the insight gained from our Industry-sponsored studies (i.e., Upstream Turnaround Performance Benchmarking Study, Oil Sands Turnaround Cost Performance Benchmarking Study), we offer our clients access to the largest database of turnaround metrics in the world.

AP-Networks' unique combination of on-the-ground experience, evaluative tools, and comprehensive data generates measurable results that bring predictability and competitiveness to high-risk events.

For a comprehensive list of our current clients, please see the 'Our Clients' section of our website.

For More Information

Asset Performance Networks
Leon van Hout
Dry Docking Benchmarking Project Manager
LvanHout@AP-Networks.com

Office: +31-20-486-1185
Mobile: +31-6-21244149

Orlyplein 10
Crystal Tower – 24th Floor
1043 DP, Amsterdam
Netherlands

www.AP-Networks.com

Benchmarking Code of Conduct

AP-Networks benchmarking studies are governed by a strict code of conduct. The principles that our studies follow include the following.

Legality
Adhere to all laws and regulations.

Exchange
Provide fellow sponsors with the same level of information that you request.

Confidentiality
Information exchanged is confidential to individuals and organizations.

Use
Benchmarking information is used only for its intended purpose.

Contact
Follow study procedures and respect corporate cultures.

Preparation
Commit to study objectives and timelines by being prepared and providing information in a timely manner.

Completion
Follow through with study commitments.

Understanding and Action
Treat fellow sponsors with respect and meet expectations.

AP
NETWORKS
Corporate Headquarters
3 Bethesda Metro Ctr.
Suite 925
Bethesda, MD 20814
USA
☎ +1-240-683-1001
🖷 +1-240-683-1009
Houston Area Office
455 E. Medical Ctr. Blvd.
Suite 200
Webster, Texas 77598
USA
☎ +1-281-554-8181
🖷 +1-281-554-8585
European Office
Orlyplein 10
Crystal Tower – 24th Floor
1043 DP, Amsterdam
Netherlands
☎ +31-20-486-1185
🖷 +31-20-486-9830
AP-Canada
407 2 St SW
Suite 1100
Calgary, AB T2P 2Y3
Canada
☎ +1-403-457-2737
🖷 +1-403-457-2767

Ocean Technical Services, LLC.

JOSEPH W. LOMBARDI
Marine Surveyor & Consultant
2600 South Shore Blvd, Suite # 300
League City, Texas 77573

Telephone Cellular (508)-958-1299
Fax (866)-381-2687, E-mail joe@oceantechserv.com
www.oceantechserv.com

19 January 2017

Mr. Joseph Stiglich, President
12316 Dormouse Road
San Diego, CA 92129
858-705-0760 joe@drydocktraining.com

Joe,

To follow is short bio followed by short discussion on talk to be given and wrapped up by a short paper on the issues to be discussed with photos.

Best regards,

Joseph Lombardi
Marine Surveyor

BIO

Self-employment History

1984 – Present President of Ocean Technical Services (OTS); a marine surveying and consulting firm specializing in structural surveys and damage assessments of yachts and commercial craft.

OTS also provides drydock & shipyard supervision, towing preparation oversight, and repair of commercial and historic vessels. OTS performs hull surveys and damage assessments for insurance underwriters, financial institutions and prospective buyers of yachts and commercial craft.

Employment History

1971-1977 U.S. Navy, Quartermaster/navigator aboard cruiser U.S.S. COLUMBUS (CG-12), PTF-23 an OSPREY-class PT boat at Coastal River Squadron Two and aboard the aircraft carrier U.S.S. NIMITZ. (CVN-68).

1978-1980 Museum of Fine Arts, Boston

1980-1988 Licensed wheelhouse operator of offshore tug/barge combinations and shipdocking Tugs for Boston Fuel & Transportation Co.

1988-1991 Licensed wheelhouse operator of offshore coastwise oil tankers for Boston Fuel & Transportation Co.

Licenses Held

U.S. Coast Guard License as Master & First Class Pilot of Motor & Steam Vessels and Tug/Barge Combinations of Unlimited Tonnage

U.S. Coast Guard License of Auxiliary Sail of 300 Tons

Radar Observer Unlimited

Affiliations

Member Society of Accredited Marine Surveyors (SAMS)
Member Historic Naval Ships Association (HNSA)
Member Society of Naval Architects and Marine Engineers (SNAME)
Member American Society of Testing Materials (ASTM)
Member American Welding Society (AWS)
Member American Boat & Yacht Council (ABYC)

Joseph Lombardi is Ship Preservation Chairman of the Historic Naval Ships Association (HNSA).

Joseph Lombardi has been named as an advisor to the S.S. UNITED STATES CONSERVANCY.

CONFERENCE SHORT TAKE

Mr. Lombardi, Marine Surveyor and Owner of Ocean Technical Services, LLC. of Houston, has been tasked by the Texas Parks & Wildlife (TPWD) to conduct the structural surveys and prepare the Structural Repair Specifications of battleship U.S.S. TEXAS (BB-35).

He will discuss the philosophy of saving historic ships (a shipyard owner & dockmaster's reason to go on vacation) with a particular bent on battleship U.S.S. TEXAS (BB-35) berthed at the San Jacinto Battleground Park on the Houston Ship Channel, La Porte, Texas.

Mr. Lombardi will briefly discuss the degradation and repairs. The thrust of the talk is on the ongoing discussion of the future preservation of this icon and the options presented. Four drydock alternatives have been produced with just one finalist.

He will talk on the need for better vessel documentation (and the challenges trying to locate this data) with specific areas such as drawings, vessel's information books and previous drydock surveys when taking on assignments for docking ships.

Mr. Lombardi will briefly discuss the recent loss of floating drydock U.S.S. RELIANCE (AFDL-47), the dangers and liability as viewed by a Surveyor, Shipyard Owner or Owner's Representative.

Ocean Technical Services, LLC.

JOSEPH W. LOMBARDI
Marine Surveyor & Consultant
2600 South Shore Blvd, Suite # 300
League City, Texas 77573

Telephone Cellular (508)-958-1299
Fax (866)-381-2687, E-mail joe@oceantechserv.com
www.oceantechserv.com

19 January 2017

ISSUES WITH THE REPAIR & DRYDOCKING OF BATTLESHIP U.S.S. TEXAS

USS Texas (BB-35) - Design & Construction:

Tracing its origins to the 1908 Newport Conference, the *New York*-class of battleship was the US Navy's fifth type of dreadnought after the preceding South Carolina-, Delaware-, Florida-, and Wyoming-classes. Central among the conference's findings was the requirement for ever larger calibers of main guns.

Though discussions commenced regarding the armament of the *Florida*- and *Wyoming*-class ships, their construction advanced using the standard 12" guns.

Complicating the debate was the fact that no US dreadnought had entered service and designs were based on theory, war games, and experience with pre-dreadnought ships. In 1909, the General Board pushed forward designs for a battleship mounting 14" guns. A year later, the Bureau of Ordnance successfully tested a new gun of this size and Congress authorized the construction of two ships.

Named USS *New York* (BB-34) and USS *Texas* (BB-35), the new ships mounted ten 14" guns in five twin turrets. These were situated with two forward and two aft in superfiring arrangements while the fifth turret was placed amidships. The secondary battery consisted of twenty-one 5" guns and four 21" torpedo tubes. Propulsion for the *New York*-class ships came from fourteen Babcock & Wilcox coal-fired boilers powering vertical triple expansion steam engines. These turned two propellers and gave the vessels a speed of 21 knots. Protection for the ships came from a 12" main armor belt with 6.5" covering the vessels' casemates.

Launch of U.S.S. TEXAS at Newport News Shipbuilding & Drydock Co., 1912.

Construction of *Texas* was assigned to the Newport News Shipbuilding Company and work began on April 17, 1911, five months before *New York* was laid down in Brooklyn. Moving forward over the next thirteen months, the battleship entered the water on May 18, 1912, with Claudia Lyon, daughter of Colonel Cecil Lyon of Texas, serving as sponsor. Twenty-two months later, *Texas* entered service on March 12, 1914, with Captain Albert W. Grant in command. Commissioned a month earlier than *New York*, some initial confusion arose regarding the name of the class.

U.S.S. TEXAS conducting speed trials off Rockland, Maine, 1914.

USS Texas (BB-35) - Early Service

Departing Norfolk, *Texas* steamed for New York where its fire control equipment was installed. In May the new battleship moved south to support operations during the American occupation of Veracruz. Remaining in Mexican waters for two months, *Texas* briefly returned to New York in August before commencing routine operations with the Atlantic Fleet. In October, the battleship again arrived off the Mexican coast before proceeding to Galveston, TX where it received a set of silver from Texas Governor Oscar Colquitt. After a period in the yard at New York around the turn of the year, *Texas* rejoined the Atlantic Fleet. On May 25, the battleship, along with USS *Louisiana* (BB-19) and USS *Michigan* (BB-27), rendered aid to the stricken Holland-America liner *Ryndam* which had been rammed by another vessel. Through 1916, *Texas* moved through a routine training cycle before receiving two 3" anti-aircraft guns as well as directors and rangefinders for its main battery.

USS *Texas* (BB-35) - World War I:
In the York River when the US entered World War I in April1917, *Texas* remained in the Chesapeake until August conducting training. After an overhaul at New York, the battleship moved up Long Island Sound and on the night of September 27 ran hard aground on Block Island. Pulled free three days later, *Texas* returned to New York for repairs.

As a result, it was unable to sail in November with Rear Admiral Hugh Rodman's Battleship Division 9 which departed to reinforce Admiral Sir David Beatty's British Grand Fleet at Scapa Flow. Finally crossing the Atlantic in January 1918, *Texas* reinforced Rodman's force which was operating as the 6th Battle Squadron. While abroad, the battleship largely aided in protecting convoys in the North Sea. With the end of the conflict in November, *Texas* joined the fleet in escorting the German High Seas Fleet into internment. The following month, the American battleship steamed south to escort President Woodrow Wilson, aboard the liner SS *George Washington*, into Brest, France as he traveled to the peace conference at Versailles.

USS Texas (BB-35) - Interwar Years:

Returning to home waters, *Texas* resumed peacetime operations with the Atlantic Fleet. On March 10, 1919, Lieutenant Edward McDonnell became the first man to fly an aircraft off an American battleship when he launched his Sopwith Camel from one of *Texas*' turrets. Later that year, the battleship's commander, Captain Nathan C. Twining, employed aircraft to spot for the vessel's main battery. Findings

U.S.S. TEXAS at anchor just after WW I.

from these efforts supported the theory that air spotting was far superior to shipboard spotting and led to floatplanes being placed aboard American battleships and cruisers. In May, *Texas* acted as plane guard for a group of US Navy Curtiss NC aircraft that were attempting a trans-Atlantic flight.

That July, *Texas* transferred to the Pacific to begin a five-year assignment with the Pacific Fleet. Returning to the Atlantic in 1924, the battleship entered Norfolk Navy Yard the following year for a major modernization. This saw the replacement of the ship's cage masts with tripod masts, installation of new oil-fired Bureau Express boilers, additions to the anti-aircraft armament, and placing of new fire control equipment. Completed in November 1926, *Texas* was named flagship of the US Fleet and commenced operations along the East Coast. In 1928, the battleship transported President Calvin Coolidge to Panama for the Pan-American Conference and then proceeded into the Pacific for maneuvers off Hawaii.

Following an overhaul at New York in 1929, *Texas* spent the next seven years moving through routine deployments in the Atlantic and Pacific. Made flagship of the Training Detachment in 1937, it held this role for a year until becoming flagship of the Atlantic Squadron. During this period, much *Texas*' operations centered on training activities including serving as a platform for midshipmen cruises for the US Naval Academy. In December 1938, the battleship entered the yard for installation of the experimental RCA CXZ radar system. With the beginning of World War II in Europe, *Texas* received an assignment to the Neutrality Patrol to aid in safeguarding the western sea lanes from German submarines. It then began escorting convoys of Lend-Lease material to the Allied nations. Made flagship of Admiral Ernest J. King's Atlantic Fleet in February 1941, *Texas* saw its radar systems upgraded to the new RCA CXAM-1 system later that year.

USS Texas (BB-35) - World War II:

At Casco Bay, ME on December 7 when the Japanese attacked Pearl Harbor, *Texas* remained in the North Atlantic until March when it entered the yard. While there, its secondary armament was reduced while additional anti-aircraft guns were installed. Returning to active duty, the battleship resumed convoy escort duty until the fall of 1942. On November 8, *Texas* arrived off Port Lyautey, Morocco where it provided fire support for Allied forces during the Operation Torch landings. It remained in action until November 11 and then returned to the United States. Reassigned to convoy duty, *Texas* continued in this role until April 1944.

Remaining in British waters, *Texas* commenced training to support the planned invasion of Normandy. Sailing on June 3, the battleship pummeled targets around Omaha Beach and Pointe du Hoc three days later. Providing intense naval gunfire support to Allied troops hitting the beaches, *Texas* fired on enemy positions throughout the day. The battleship remained off the Norman coast until June 18 with its only departure a short run to Plymouth to rearm. Later that month, on June 25, *Texas*, USS *Arkansas* (BB-33), and U.S.S. NEVADA and USS *Nevada* (BB-36) attacked German positions around Cherbourg. In exchanging fire with enemy batteries, Texas sustained a shell hit which caused eleven casualties. Following repairs at Plymouth the battleship began training for the invasion of southern France.

U.S.S. TEXAS firing at targets ashore, Cherbourg, France, 1944.

After shifting to the Mediterranean in July, *Texas* approached the French coast on August 15. Providing fire support for the Operation Dragoon landings, the battleship struck targets until Allied troops advanced beyond range of its guns. Withdrawing on August 17, *Texas* sailed for Palermo before later departing for New York. Arriving in mid-September, the battleship entered the yard for a brief overhaul. Ordered to the Pacific, *Texas* sailed in November and touched in California before reaching Pearl Harbor the following month. Pressing on to Ulithi, the battleship joined Allied forces and took part in the Battle of Iwo Jima in February 1945. Leaving Iwo Jima on March 7, *Texas* returned to Ulithi to prepare for the invasion of Okinawa. Attacking Okinawa on March 26, the battleship pounded targets for six days before the landings on April 1. Once the troops were ashore, *Texas* stayed in the area until mid-May providing fire support.

U.S.S. TEXAS, final wartime appearance.

USS Texas (BB-35) - Final Actions:

Retiring to the Philippines, *Texas* was there when the war ended on August 15. Returning to Okinawa, it remained there into September before embarking American troops for home as part of Operation Magic Carpet. Continuing in this mission through December, *Texas* then sailed for Norfolk to prepare for deactivation. Taken to Baltimore, the battleship entered reserve status on June 18, 1946. The following year, the Texas Legislature created the Battleship *Texas* Commission with the goal of preserving the ship as a museum.

Raising the necessary funds, the Commission had *Texas* towed to the Houston Ship Channel near the San Jacinto Monument. Made flagship of the Texas Navy, the battleship remains open as a museum ship. *Texas* was formally decommissioned on April 21, 1948.[i]

U.S.S. TEXAS at San Jacinto Battleground Park.

CURRENT SITUATION

The battleship has been berthed at San Jacinto since 1948 and was, initially, flooded with brackish water to allow her to settle into her berth. In 1988 she was towed to drydock at Galveston, almost sinking on the way. Her flooding during the transit to the drydock only allowed less than 1' of clearance over the keel blocks at time of docking. Repairs were carried out with much doubler plate being added to her shell plating.

U.S.S. TEXAS in drydock, Galveston, 1989.

It is approaching 30 years since the ship was drydocked and had what can be considered 'Band Aid' repairs that were not designed, given the high galvanic potential in the Houston Ship Channel with no cathodic protection, to last more than the life cycle of the paint system and doubler plating (15 years).

As a result of the previous degradation to the ship as a result of the flooding and lack of budget to put things really right the ship began a series of flooding events due to very thin shell plating almost immediately after re-berthing at San Jacinto.

Cathodic protection was added and four mooring clusters were fitted to allow the ship to be pumped out and be properly secured.

In June of 2012 a major flooding event induced progressive flooding throughout the area aft of the boiler rooms. Repairs were undertaken to make most of the transverse bulkheads watertight to mitigate the effects of progressive flooding. Additional pumping capacity was added to allow the ship's company to arrest any major breaches. Phase I Repairs were completed in early 2015.

Phase I repairs were developed to arrest the most serious leaks and to rebuild the tank structure and watertight bulkheads in the Engine Room, tank spaces aft of the Engineroom and a large void D-12 at Frame 122 - 129.

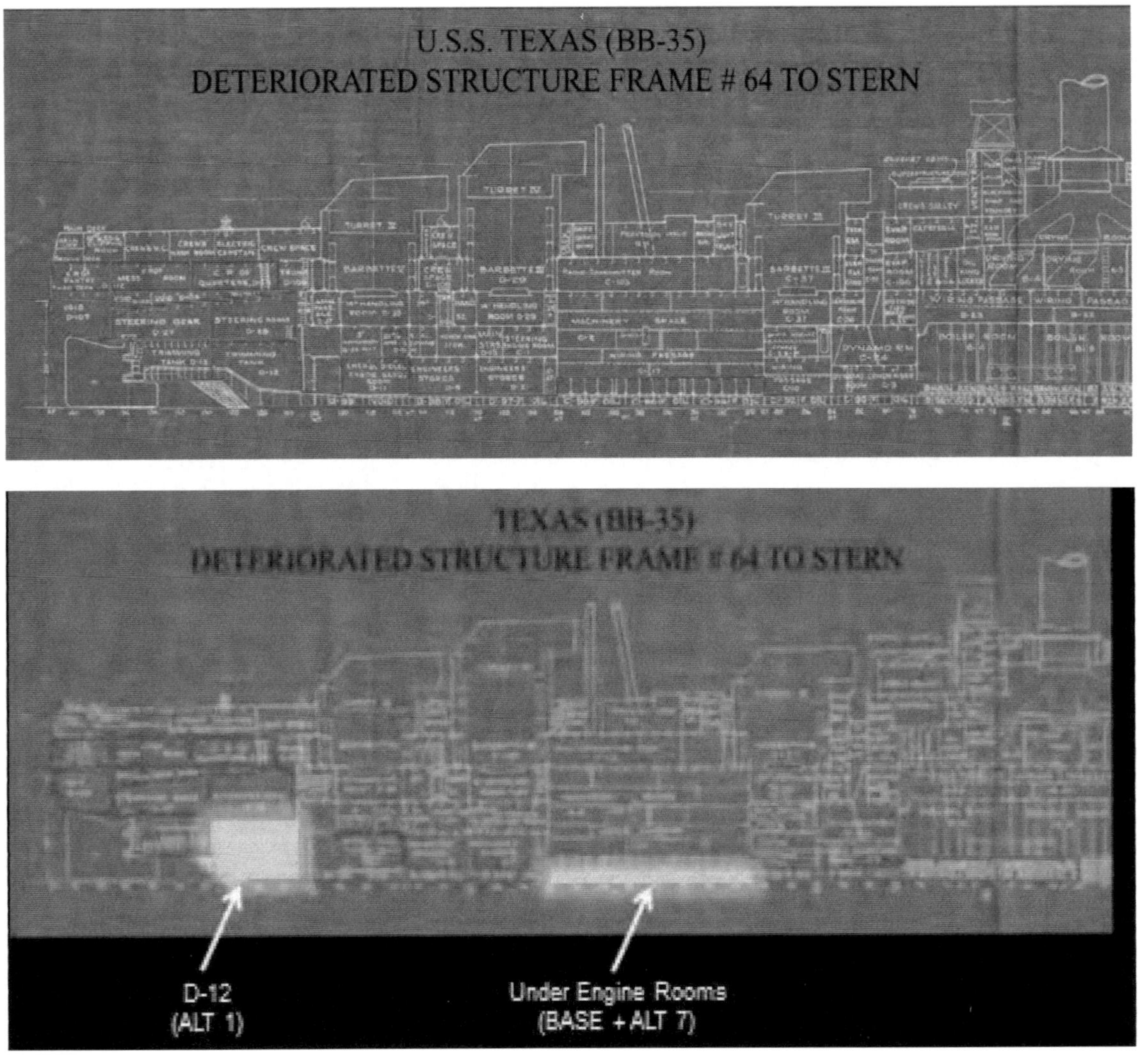

PHASE I Repairs.

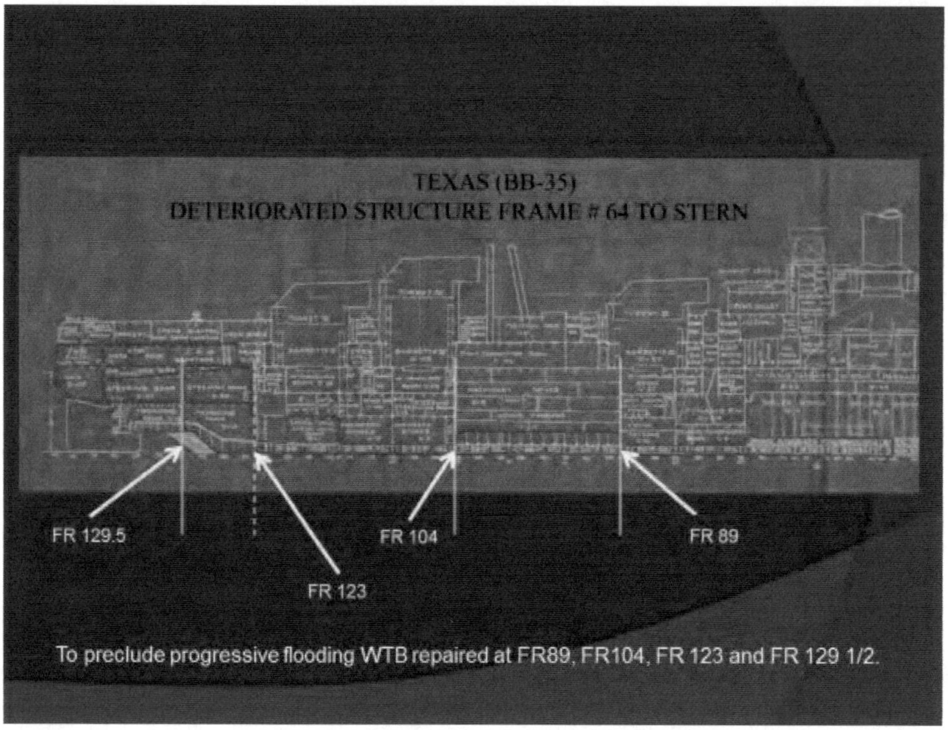

Both steam engines were supported from the armored 3rd deck to allow the repairs to be made to the Inner Bottom tankage.

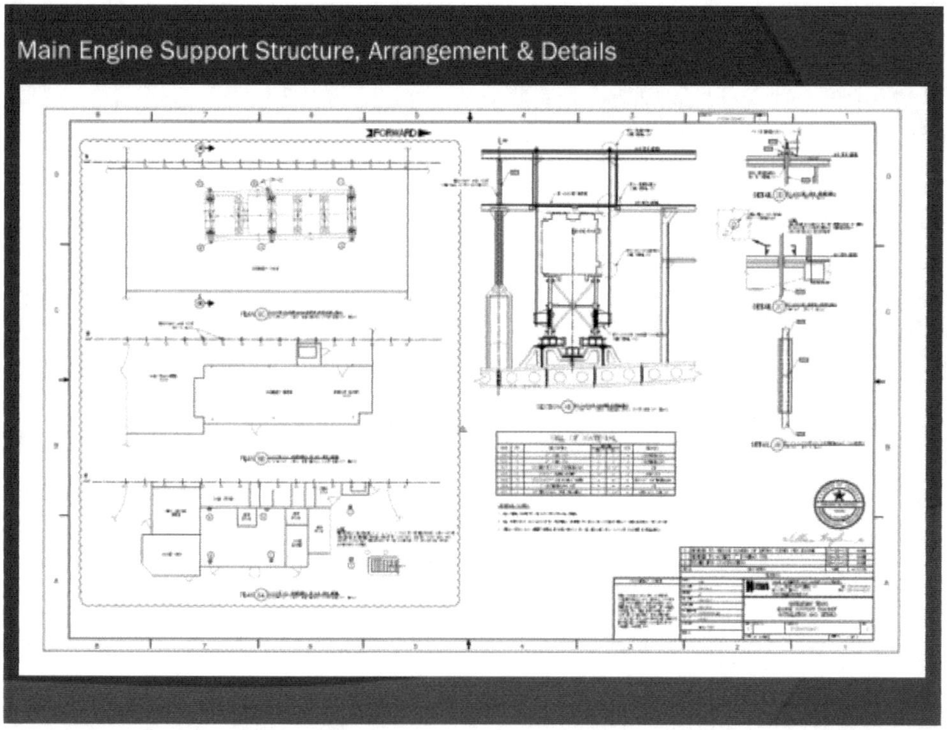

Bottom photo shows totally wasted transverse frames portside engineroom; top photo shows subsequent insertion of sistered 3/8" steel plate.

Wasted framing in D-12 Void.

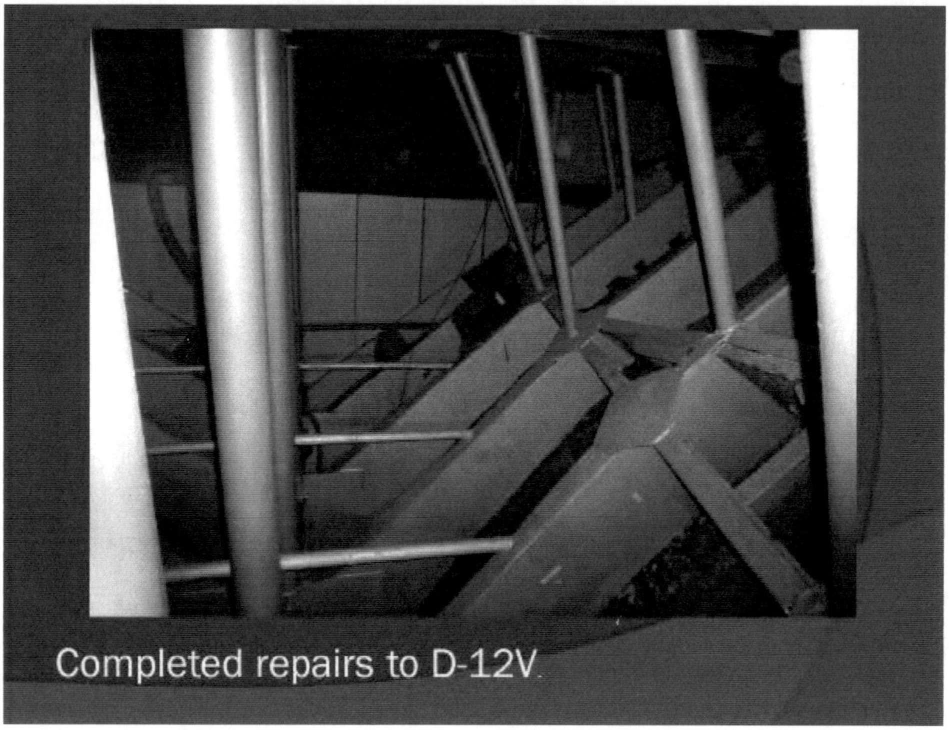

As part of the work package, TPWD tasked AECOM with designing three (3) drydock concepts to allow the ship to be kept at San Jacinto. The stake holders and the U.S. Coast Guard would not allow the ship to transit the Houston Ship Channel given her precarious condition. The four concepts were presented as depicted below.

The above concept drawing was chosen and will be used to construct a drydock for the U.S.S. TEXAS upon completion of Phase II repairs and additional funding.

U.S.S. RELIANCE (AFDL-47) FLOATING DRYDOCK

As a surveyor, ship owner and drydock operator we all come in contact with various types of equipment throughout our careers. Some drydock facilities are more challengine than others, but all require a close level of scrutiny as to their material condoition and operating systems.

In this country the industry is still largely relying on WW II or earlier graving docks or floating drydocks. The owner, underwriter or surveyor has an approaching challenge for ensuring his client's vessel is safely berthed for repairs.

In my experience, I have seen floating drydocks sink from lack of maintenance (AFDL-47) and had experience with leaking graving docks (Drydock #1) in Charlestown, MA. The AFDL-47 sank the morning that I had a destroyer (U.S.S. LAFFEY) under tow for Detyens Shipyard in N. Charleston, a very challenging time as the destroyer had wasted keel, transverse & longitudinal frames and pourous bulkhead that had recently been the subject of progressive flooding.

We docked the ship pierside for over a week while the yard scrambled to replace electrical equipment; this was done and a test was made of that drydock. She sank again and another round of repairs were required.

As a surveyor, I insisted that the yard lower and raise the drydock several times upon completion of the repairs to ensure my vessel would not be harmed. The resulting drydocking was uneventful and the ship was superbly repaired.

U.S.S. LAFFEY (DD-724) in floating drydock AFDL-47 for repairs.

Years went by and this past September I received a call from underwriters to conduct a Trip-in-Tow and Tow Plan for the floating drydock AFDL-47 as she was to be purchased from Detyens by a ship recycling firm at the old Roosevelt Roads Naval Station in Puerto Rico. I had surveyed this drydock in 2010 when Detyens decided to purchase the drydock from the Navy.

Surveyors are asked to do many things and they are usually in the middle of two or more parties and must answer to many different masters. They are also responsible for their actions and reports. I reported what I had seen to the underwriter and advised that the drydock needed extensive repairs; their conservative approach was that they would cover 'port Risk' while the drydock was pierside, but they would not bind coverage for the tow.

The Owner of this drydock did not have the drydock surveyed before purchase. I asked that the drydock be pumped out, which was done, and I began my inspection of the forward rake tankage. I had finished inspecting the forward rake and and the new Owner stated to me that he could not get hull/liability coverage for the drydock for the upcoming trip. He stated that his company was going to self insure the drydock. They then proceeded to load about $ 1M in scrapyard machinery and plate steel on the pontoon deck for the tow. I pointed out to him that both rake ends were detached (deck from vertical rake) and that there were hundreds of holes on the pontoon deck and on the wing walls that needed plating. I also stated that I had only just completed inspection of the two forward rake compartments.

I ended my further participation in the survey.

As the Owner was self-insuring the drydock I decided that thius was not a good situation and I ended my participation in this work. I went back to the yard a week later and the 1 man crew has squirted foam into the holes on the deck.

AFDL-47 was taken in tow for Puerto Rico and sank 260 miles SE of Jacksonville, Florida 3 weeks later.

[i] http://militaryhistory.about.com/od/usnavy-battleships/fl/World-War-III-USS-Texas-BB-35.htm

Acquisition and Modification of HMB-1

by Glen Roberts

My presentation's topic is Bay Ship and Yacht's Acquisition and Modification of the HMB-1. I'll primarily focus on the recent history, but will also briefly discuss its remarkable origins.

HMB-1

To get an idea of what I'll be talking about, let's take a quick look at the particulars of the HMB-1 (short for Hughes Mining Barge). A stripped-down arrangement drawing of the HMB shows a somewhat unique shape, with a fairly small length to beam ratio. It's 324' long overall, with 106'8" extreme beam and 76'8" wingwall opening. The wingwalls are also tall at 43'3" above the main deck. It's designed lift capacity is 6290 LT. Also notice the forward end is closed with a bulkhead, as well as the telescoping roof. Detail on the structures and equipment will come later, as part of Bay Ship's modifications.

PROJECT AZORIAN

Now that we know what the HMB looks like, let's see the reason for its design. It involves the story of the Glomar Explorer, part of code name "Project Azorian". I'm sure most readers will have some knowledge of this, so I'll quickly highlight what I found to be the most outstanding facts, particularly the scale of the engineering involved.

The story begins in 1968 when the Soviet K-129 sinks north of Hawaii. The K-129 was a diesel-electric ballistic-missile submarine. The Soviets scoured the general area for floating debris, with only a vague idea of where to look. They looked for months, but were ultimately unsuccessful.

The United States had sonar sensors positioned in the Pacific to monitor nuclear testing. They detected the K-129 casualty and were able to triangulate its position to within a few square miles. The remains were soon located and photographed in the 17,000 foot water with remotely operated submarines, which were also a unique capability of the US.

The sunken Soviet submarine was believed to have carried 3 nuclear missiles and 3 nuclear torpedoes, in addition to valuable cryptography equipment. This, along with the Soviets being in the dark, proves too tantalizing a prospect for the US to ignore. A decision is soon made to attempt a clandestine salvage operation.

About six years lapse pass between the sinking and retrieval. A quick summary of key events include:

- 10/70 – Pipe string lifting concept is chosen.
- 8/71 – Project almost canceled, is given 10% expected chance of success.
- 11/71 – Glomar Explorer keel laid at Sun Shipbuilding in PA. Ship delivered 4/73.

- 6/74 – Nixon approves mission on the condition it not start until after his Moscow summit.
- 7/4/74 – Glomar arrives at recovery site, one day after Nixon's summit.
- 8/1/74 – Lift operation begins
- 8/8/74 – Sub remains secured in Glomar Explorer
- 8/9/74 – Nixon resigns

Project Azorian required many components to dead-lift a large submarine from very deep water. Keeping in mind the above-water appearance must look innocent, here's a list of the main features:

- A capture vehicle (CV) lands on top of the submarine and secures it with hydraulic arms
- A pipe string lowers the CV from the Glomar Explorer, which was made of sections of threaded steel pipe
- Docking legs assist in the load transfer from the pipe string to ship
- The pipe sections were assembled and fed up & down with sophisticated pipe handling gear
- To counteract the ship's pitch & roll in the ocean, the pipe string was suspended from a derrick, which was mounted to gimbaled platform
- The Glomar Explorer itself is 619' long, 116' wide, 50,000 LT. It contained a large well space that could be closed and pumped dry.

The gimbaled platform was required to isolate ship's movement and dynamic loads from the suspended payload. A gimbal is a set of concentric bearings that ensure something lies flat, like a compass, which eliminates bending stresses in the pipe string. Each had a capacity of 5,000 tons and were the largest ever machined. The platform accommodated 8.5 degrees of roll and 5 degrees of pitch.

The gimbals were mounted atop heave compensator rams to negate the up & down bouncing of the seas. It was a hydraulic/pneumatic system in which the air pressure was adjusted to match the load on gimbaled platform. It accommodated 15 feet of heave and had a 10,000 ton capacity.

The heavy-lift pipe string was assembled in a tapered fashion, using six different diameters, from 15.5 to 12.75 inches. It was made of forged steel, with 570 joints of 30-foot sections, for a total length of 17,000 feet. The pipes were pre-assembled in doubled-up 60-foot segments, which weighed 12 to 20 tons per pair.

I won't be able to describe every aspect of the pipe handling system, but basically a pipe section was grabbed manually from below deck with an overhead gantry crane. The pipe was then loaded into a transfer cart & fed to a vertical elevator. Two pairs of hydraulic arms would pick, fasten, and release the load in a continuous, hand-over-hand sequence. The lift system was designed to raise or lower a 60-foot pipe section every 3.3 minutes (although it operated slower).

The Glomar Explorer's center well is 199' x 74', with a 65' clear vertical height. The hollow midbody and extreme loads being carried overhead required extreme design and engineering. The dry well is closed by two gates, which are really 9' tall barges. They roll longitudinally, forward and aft, driven with a rack & pinion drive system. They're guided by rails on the bottom of the ship's hull.

A watertight seal is formed by controlling the gates' buoyancy with compressed air. To seal the well, you pump air into the gates, they float up against rubber gaskets on the ship's hull, and then you drain the well. The pressure differential holds them firmly in place. The same concept keeps them in place in the open position.

The docking legs are two semi-rigid structural arms that could grab the suspended weight and also reach below the ship's hull to guide the "claw" and payload as it is hoisted into the center-well. They're needed to mate the stabilized underwater machinery to the dynamically moving ship, and then hoist it into the narrow confines of the ship's center-well. They're mounted to hinge pins that allowed them to splay out longitudinally.

The capture vehicle (code named "Clementine") is a large box girder weighing 2,000 tons. It's fitted with multiple hydraulically operated, articulated fingers. The fingers are large (some over 30 feet long) with multiple articulated joints. They're made of high strength steel plate, 1 to 2 inches thick. The hydraulic joints are driven using seawater pumped down through the pipe string. Eight propellers position the CV over the target with an acoustic positioning system, independently of the ship above. The ship then senses the bias in the pipe string and commands the ship's thrusters to position the ship directly over the CV. The claw needed to land within 2 feet of its intended position over the submarine. After the fingers have closed around the submarine, the system needed help freeing the submarine embedded in the mud. The four corner legs are hydraulic cylinders, also driven with seawater, that contribute an extra 2,000-ton lifting force. The cylinders are jettisoned behind before starting the lift.

It's a shame to conclude the Azorian story so abruptly, but in summary: the submarine was lifted about halfway, to a depth of 9,000 feet when one or more of the claw's fingers failed. They dropped 2/3 of the sub, retrieving only the forward third.

By now, the reader is probably wondering how the HMB-1 is involved. During the planning stage, it was decided that the capture vehicle be built concurrently with the Glomar Explorer. However, the CV's existence couldn't be explained away with a cover story. So it had to be constructed in secret and loaded inside the Glomar without being seen. This required a fully submersible barge that could be completely shrouded from public view.

In 1969, Lockheed contracted Glosten to design the HMB, which was completed in 15 months. NASSCO had it built 9 months later. One of the only available photos of "Clementine" shows it in the HMB-1 with the roof pulled open.

After the claw was fully outfitted inside the HMB, the pair was transported to Catalina. Here it rendezvoused with the Hughes Glomar Explorer, ready to load the capture vehicle in its well space. At night, the HMB was submerged to the bottom of the flat, shallow seabed, and its roof was opened. The Glomar then transited over the HMB and carefully lifted the claw into the center-well with the docking legs.

During operations, all pneumatic and hydraulic power was supplied by equipment on a nearby support barge. When submerged, the HMB's water plane area is at a minimum, and a very small flooding rate is required. Four buoyant cylinders at the corners provide additional stability.

To maintain control of the barge while submerged, some tanks needed to withstand the water pressure while empty. Therefore, the HMB is divided into "hard tanks" and "soft tanks". The soft tanks comprise all the compartments below the 01 deck (10' above main deck). They were open to the sea and would completely flood. They're rated for a head pressure of 42 ftsw.

The hard tanks are smaller compartments in the wing walls above the 01 deck. The section view drawing shows heavier frame material. The internal boundary test head is 177 ftsw. Two spaces near amidships are control spaces with air locks for submerged entry.

The ballasting system is pneumatic, with remotely operated valves on sea chests and manual/remote valves for the vent blow system. Sea chest for the soft tanks are located just inboard of the wing wall with each chest serving a pontoon and wing tank. Hard tanks receive or discharge water through sea chests just below the 01 deck on the inboard side of the wing walls.

A list of the HMB's systems shows it was a complex piece of machinery:

- Ballast Air Supply and Breakout system
- Constant Pressure Air Supply and Overpressure Protection System
- Hydraulic System
- Wet Lock System
- Control Space Pressurizing, Venting, and Atm. Reference System
- Soft Tank Flooding and Blowing
- Variable Tank Flooding and Blowing
- Trim Transfer and Filling System
- Stabilizing Cylinder Flooding and Blowing
- WB Reel and Winch Power and Control
- Sea Water Cooling System
- CB Winches, Power, and Controls
- Electrical System

After Project Azorian, the HMB sat unused for years in the Bay Area until the mid-1980s when it found another use. The Sea Shadow, another top secret project, was built completely inside the HMB. Along with stealth technology, the program explored the possibilities of modular construction where various chunks of the ship were built by different fabricators who presumably did not know what the finished product would look like.

After the Sea Shadow was built, the HMB served as a "bat cave", or wet berth that could be closed off to prying eyes. A roll-up "mega door" and bell mouth structure were installed on the HMB's for the Sea Shadow. During testing, the HMB with Sea Shadow inside was towed out to sea, and the Sea Shadow could trial under the cover of darkness. In 1993 the Sea Shadow was announced to the public and until 2006 was used for daylight testing, war gaming and PR work. It was then anchored with the Suisun Bay ghost fleet.

BAY SHIP & YACHT

ACQUISITION

That brings us to the latest chapter in the HMB's story, which is Bay Ship & Yacht's acquisition. The first question to ask is: why the HMB? Obviously, it was an additional drydock facility, which allowed us to dock two large (for us) vessels at a time. But what additional capabilities did it offer us?

The next biggest reason was to handle wider-beamed vessels. We were previously limited to vessels 60' wide or less, although even 60' was a stretch. Our WW2-era AFDL drydock has sloped wingwalls, 57' at the bottom and 64' at the top. A 60-foot wide barge required a 14' docking height. Although this made it easy to center the barge, it only allowed 5-1/2" clearance at the bottom corners. This made for a very labor-intensive build, including: stacking the cement blocks, welding steel braces to the block tops, and using most of blocks in yard. It was also much more difficult to sandblast, paint, and inspect.

The HMB would allow us to dock boats up to 76' wide, principally the barge market. Some other reasons included:

- Maximum lift capacity more than doubled, allowing for bigger boats, or not having to de-ballast before docking.
- High wing walls allow us to go 17' deeper. It will ground out in the shipping channel before the wing walls go under. This allows deeper draft vessels, taller block heights, and less required de-ballasting.
- A steel dock offers more blocking options compared to concrete, able to weld to the deck
- The retractable roof was attractive for helping with protecting against the elements and environmental containment
- 15 ton overhead gantry crane
- Although our AFDL drydock is still going strong at over 70 years old, we don't expect it to last forever.

With the HMB's potential value being established, it was time to take a first-hand look. BSY employees made several trips to the Suisun Bay, starting in 2008. She had been sitting there for about two years by this time. This was to evaluate the overall condition, see what equipment was on board, what modifications were needed, take photos, etc.

After being mothballed for several years, some items features were noticeably corroded, but the important thing was the structural integrity. We did an ultrasound survey in 2010 that showed barely any deterioration of the ½" steel shell plate.

With the HMB appearing to be in good shape, Bay Ship was ready to negotiate the terms of purchase. The US Government's original offer was to donate the HMB-1 and Sea Shadow as a package deal with the idea that the Sea Shadow would become a museum piece that visitors could tour. We drew up plans showing the Sea Shadow in a corner of Bay Ship's yard, just south of the Syncrolift. We attempted to team with the Alameda Hornet Museum, but the museum didn't meet the qualification requirements.

There were no other takers, so the US Government auctioned the HMB-1/Sea Shadow in an online bid. Bay Ship was the second highest bidder behind a consortium of Chinese investors represented by US Law Office. They were disqualified due to sensitivity of the stealth technology. Bay Ship was offered the package and accepted. With deed in hand, Bay Ship towed the HMB from Suisun Bay to our pier at Treasure Island to begin the outfitting process in July, 2012.

However, the purchase agreement had a requirement to take care of first. The auction stipulated that the Sea Shadow had to be scrapped in the US under Navy Monitoring Program. Bay Ship contracted Zaccor to scrap the Sea Shadow. They used two back hoes equipped with hydraulic shears. The contract mandated that it be scrapped down to pieces of a certain size, roughly 3-4 feet.

MODIFICATION

With the deck clear, it was time to modify the HMB to make it operational. We first had to decide how we wanted control it as a drydock— essentially what tanks to use for ballasting.

During the bidding process, we were given the original operating manual written by Glosten. It was 116 pages, and the word "caution" appeared at least 100 times. Another item provided was a survey done during a 1992 drydocking. It mentioned "blown" bulkheads that were repaired at a cost of around $1.3 million.

It was an intriguing idea to keep using the compressed air ballasting system, but if sophisticated company like Lockheed could accidentally blow bulkheads during docking operations, we thought it would be prudent to simplify things. We wanted to mimic the operation of our existing drydock, which has 8 ballast tanks. We took it a step further and decided to go with four tanks. This cuts down on machinery, maintenance, simplifies operation. Also, using just four corner tanks almost eliminates the possibility of putting large hog or sag bending moments in the dock.

The four large saltwater ballast tanks in the corners were created by cutting holes between existing adjacent tanks. In addition, we use 4 permanently filled freshwater ballast tanks at the rakes and middle tanks. On occasion we've had to empty these to get extra buyoancy for heavier vessels. Tanks are connected longitudinally with crossover valves creating redundant pump & valve control.

The pumps and valves are located amidships, adjacent to the middle permanent ballast tanks. The tanks are emptied with 5000 GPM electric pumps. The 14" gate valves are operated by hand cranks at the top of wingwall.

For controls & instrumentation, our systems engineer designed an elaborate console. With cost control in mind, he used off-the shelf parts from Automation Direct. The pumps can be operated locally at the motor control centers at the midships hutch, from the main control house, or via handheld wireless remote.

The control house console includes touchscreen control, valve position indicators, tank level indicators, and digital list and trim readouts. The interface is highly programmable: pumps can be set to run until a certain tank level is reached; alarms can be programmed to sound at preset tank levels or list/trim angles (intended for leak detection). An auto-dialer will forward a pre-recorded message to a contact list. The design got recognized in "Automation Notebook", a trade magazine published by Automation Direct.

Being in the Bay Area, we not only had to worry about the HMB itself, but also the environment around it. The Department of Fish & Game was concerned about delta smelt being sucked through the flooding valves, and advised that we design and use screens to protect the fish. Four screens were made to cover each of the flooding valves. They're normally stowed above water, and lowered down over the side with an electric winch for docking operations. Inside the screen is a ¼" steel baffle plate with non-uniform holes to diffuse the water flow.

There was also concern that fish might swim into the drydock well and get trapped. So we installed a bubbler curtain along the dock sill which is connected to the yard air supply. The bubbles confuse the fish that then swim the other way.

After 10 months at Treasure Island, and most of the major upgrades in place, the HMB was towed to the main yard in Alameda in May 2013. Its pier space is adjacent to the Alameda ferry terminal; having such a wide barge next to the ferry landing made it difficult for the ferry captains to pull away from the float and into the estuary. Being good neighbors, we helped them shift their float about 107' to the west, away from the HMB. We also re-routed their pedestrian catwalk and installed new gangway support structures for the various ferries used.

The Bay Area has an ongoing project called the Bay Trail, a pedestrian path circling the waterfront. We worked with BCDC (Bay Conservation and Development Commission) to improve it by adding pavement and landscaping.

With all the hardware in place, we just had to get our paperwork in order. The Coast Guard Certification included: an underwater survey by Ballard Marine Construction that checked the seachests, zincs, signs of hull damage; a visual tank inspection with marine surveyors, checking the hull plating and stiffener condition; review and stamp by a professional engineer

DOCKINGS

This report will conclude with some of our noteworthy dockings. We docked our first barge in June 2013, cryptically named BAL0010 or the CIB723. It might be more familiar as one of the "Google Barges" that was the subject of much media speculation. The 62' wide barge had plenty of clearance at the sides.

The Manson DB Valkyrie, a derrick barge, was our fourth docking in August 2013. At 72' wide, it also made good use of the increased beam capacity. The large crane was more difficult to fit under the roof, and had about 5' of clearance.

After docking a series of 10 barges, it was time for something a little more interesting. In October 2013 we got the plans for the sailing yacht Athena. A 3-masted schooner, 295' x 40', with the primary blocking area is under its deep skeg. However, the captain required support under the forward and aft masts. Even with the skeg blocked low on 12" blocks, the taller block support pyramids reach 11' above deck. This meant we had to float over this block pyramid, which required 30' of water over the HMB deck, or 49' feet total.

Sinking so deep required opening upper tank covers on the wingwall to let the upper tanks flood, increasing dive depth. The tall side blocks were held in place with angle braces welded to doublers on deck.

After several more barges and a LCU, the 15th docking was Scripps research vessel Melville, 280' x 46'. The Melville was memorable because the customer requested a keel height increase from 4' to 7'8" three days before docking. This was a critical issue because of starboard and stern A-frame davits needing to fit under the roof. I scrambled to find drawings of deck davits, and estimated about 20" clearance under the roof section. This was a little too close for comfort, so I pushed the boat forward a few feet, which left the bow tip hanging about 25' over the dock sill.

In March 2015, the Ross Chouest arrived for an emergency docking. At 256' x 54', and displacing 4220 LT, it was 50% heavier than the USCG Morgenthau class, our previous heaviest docking. We had to partially de-ballast vessel to get the system stability to a conservative 5' of GM.

Damage Repairs To A Composite Steel/Concrete Floating Dry Dock While In Transit Aboard A Heavy Lift Vessel

Presented By Wally Sayed
Heger Dry Dock, Inc.
2017

My Mediterranean Adventure begins in Portland, OR. While conducting a routine dock inspection in Portland, a Mediterranean client called to schedule their inspection of their newly purchased used dock as soon as possible. With one small caveat; It is in transit on a heavy lift vessel (HLV) and was damaged during the initial lift. Preliminary indications are that that the cribbing has punched at least two holes into the bottom shell.

Did I mention that the dock, built by Pallada in Ukraine for the Russian Navy in 1973, is a composite dock with steel wings and a reinforced concrete pontoon?

With my inspection in Portland concluded and 76 hours at home, I left Boston with a one-way ticket to Gibraltar. Long distance arrangements had been made for me to meet a three-man repair team from the shipyard in Gibraltar where the HLV was due to pass. With only 12 hours in Gibraltar, the four of us boarded a Pilot boat in the middle of the night and headed out two miles offshore to board the heavy lift vessel for the remaining 5 day transit. The five days were intended for us to assess the damage, determine the materials required and prepare and coordinate all repairs to be made on arrival but prior to float-off. Early on, it was determined that the float-off would occur 24-hours after arriving at our destination in order to affect all repairs. While some supplies were brought on board with us, such as plywood and saws, much of the remaining patch repair components were not available until we reached our destination.

Weeks prior, we had been contacted to provide inspection and certification services to an overseas client regarding the composite dock. Our drawing request yielded enough information, in Russian, to assess the docks buoyant capacity. The reinforced concrete details, while detailed for the concrete dimensions, did not include the steel reinforcement details. It did indicate that the pontoon deck and bottom shell are 6-inches thick, and the watertight and –non-watertight bulkheads are 4-inches thick. The strength must be in the rebar. On my way to Gibraltar, I was hoping to find the remaining details aboard the dock.

Prior to departure, a basic repair plan was put into place.

1) Using the general arrangement, develop a GHS model and replicate the damaged tanks to assess damaged stability, drafts, list and trim.
2) Determine a viable method of temporary concrete repairs that can withstand up to 10-feet of hydrostatic head pressure.
3) Determine a viable method of permanent concrete repairs that can withstand up to 30-feet of hydrostatic head pressure.

1) Based on our knowledge that Tanks 10 and 11 were breached, a GHS model was run to replicate the flooded tanks. If no repairs were made, the dock would experience a xxx-inch trim and a xxx-inch list with a minimum of xxx-

inches of freeboard. That is, the dock will float crooked and the pontoon deck will remain dry. This foreknowledge was enough to set us all at ease; Even if the repairs do not work as intended, the dock will not sink.

(You may be asking yourself, "If it won't sink, why make the repairs now?". To which there is a very political answer. Upon delivery, the dock will transit under tow several miles from the Mediterranean, through a Grand Harbor to the shipyard with a cast of unwitting thousands watching, potentially including political figures, competing interests and journalists, and the newly arrived dock cannot be seen to be limping home.)

2) We contacted Southern Concrete and Jim Tombarello, concrete repair experts, to assist with generating a repair plan. We have successfully worked with both individuals in the past making repairs to commercially operated AFDM's which are 100% reinforced concrete docks. Discussions with them led us to Green Mountain grout products which manufactures specialized grouts for concrete repairs. Product review revealed two possible methods: 1) Epoxy based and 2) Expanding foam. While the epoxy base would provide higher strength, it would require skilled hands to mix and place as well as requiring adequate ventilation. Instead, Gel Foam 2 was chosen as the grout of choice. The Gel Foam has an initial viscosity similar to motor oil and is hydrophilic, or water absorbing, and does not require ventilation. Its intended purpose is to seal cracks in underground utility tunnels or dams where there is active seepage under pressure. The crack is prepped and cleaned and the foam is applied into the crack. The seepage reacts with the gel base creating a foam that thickens and expands to approximately 6 times its original thickness effectively sealing the crack.

3) Initially, the long term repair plan was to construct a timber cofferdam fitted underwater in way of the damage, remove the damaged concrete, clean the rebar (we all hoped was there) and re-pour the bottom shell using the cofferdam as a form. That was before I inspected the damage.

The bottom damage was initially discovered when the dock was lifted and muddy water was observed running off the bottom of the dock onto the HLV deck. The marine surveyor inspected tanks in way of the external leak point and found two obvious fractures in the bottom shell in Center Tank 10. The adjacent starboard Wing Tank 11 was suspect, but no obvious fracture was found.

When I arrived on board the heavy lift vessel (HLV), I received the cribbing plan for the dock. The cribbing arrangement was laid out in a herring bone pattern using a single layer of 12"x12" (300mmx300mm) hardwood timbers clipped to the deck of the HLV. The pattern was intended to pick up multiple frames with a single timber and alleviate the risk of missing the longitudinal offset trying landing on such narrow frames. While well intended, the unintended consequence was the damage that followed.

With the cribbing plan in hand, a cursory external inspection of the dock was performed to determine any root cause to the damage. While not necessarily within the scope of work, it was important to make this determination in the event that the cause was still in effect.

It was quickly determined that the outer edges of the dock were unsupported by the cribbing, and it became apparent that there was camber, or deadrise, to the docks bottom shell. This was not noted on the general arrangement plan we received. However, the floating dock had recently been in a graving dock, so information must have been available then to shape the keel blocks to match the camber (or they would have damaged the dock then). Further inspection revealed that the dock was visibly flexing on the blocks, demonstrating cantilever bending. The HLV captain was immediately notified and a strong recommendation made to add cribbing blocks at the edges in order to reduce the amount of flexure.

Support against an un-stiffened panel is not recommended. The frames of the dock are rigid in comparison to the un-stiffened bottom panels in between the frames. Each frame the cribbing crosses is a narrow point load bearing on the cribbing. This small patch is the effective area, not the entire length of the cribbing. As the load is transferred from the frame to the cribbing, a local amount of compression to the cribbing occurs at the frame, with the remaining section of cribbing bearing on the un-stiffened panel. The cribbing spanning between the frames is not loaded by the panel and is uncompressed, thus higher, causing it to "push" on the panel. This condition generates large shear forces adjacent to the load point that must be supported by the panel. Or not.

In addition, the hardwood will compress less than soft wood as it picks up an increasing load. As such, it is unable to make up for any large deformations in the hull which causes the reduced area in contact to take up the full load instead of being distributed as intended. The reduced area becomes a hard point and failure is based on the weakest link. The block may punch through the deck below, or the block may fracture and fail, or the block may punch through the hull of the lifted vessel. Once failure has occurred, the load will redistribute to the remaining blocks, which must now carry a load greater than anticipated. Subsequent failures may continue to occur if the redistribution causes other points of overload and fail i.e. the domino effect.

Generally, accidents are not due to a single erroneous event, but multiple errors that compound themselves into unforeseeable events. This accident is no exception. The cribbing inspection revealed that the cause of the damage was still in effect. The effective cribbing area was significantly reduced by lack of contact and the unsupported panel would have to carry a much higher load than intended or designed. Typically, the bottom panel is designed to withstand hydrostatic pressure only; It is not intended to be load bearing. Additionally, the dock was flexing at the edges where it was unsupported.

The short cribbing made for a narrow clearance that made it impossible to inspect the bottom shell. While increasing the cribbing height would increase the VCG of the dock on the HLV, discussions with the HLV captain indicated there was sufficient GM in the HLV that stability was not an issue. The HLV was designed to transport rigs with a very high VCG. In hindsight, it would have been feasible to design the cribbing 24-inches high, while maintaining a 2:1 aspect ratio.

With the cribbing inspection complete, the internal dock structure was inspected starting with Tank 10. Immediately upon entering the tank, the damage to Frame 88 was obvious and alarming. When the area was cleared away, the bottom panel had a compound fracture extending 15-ft transversely by 8-ft longitudinally, which included fracturing the frame itself in way of the lightening hole. The transverse crack stopped and ran longitudinally along the edge of the watertight bulkhead separating Tank 11. Further inspection of Tank 10 also found a shear fracture at Frame 70, which ran 15-ft transversely along the edge of the frame. The alarming sight of the crack opening and closing further reinforced my panel shear theory. This observation led me to suspect that the potential for this type of failure could occur in any tank, and so the 100% structural inspection began.

Going to Tank 11 next, the suspected hole was searched for but not found. The watertight bulkhead in way of Frame 88 was cleared and no damage was found. In the remaining tanks, there were indications of excessive loading, but no further bottom damage was found. Tank #7 had a set of fractures in way of the opening on Frame xx, but the presence of rust film indicated this was not recent damage, and the crack did not go through the bottom. Other frames had localized spalling of a thin sheet of concrete cover in way of a cold seam near the pontoon bottom.

During the internal inspection, a curious component of the dock deballasting system was found. All of the dock deballasting pipes had a small pipe attached to the top of the bellmouth, many of which were broken off. I was later informed that this was a steam line to defrost the ballast pipe. However, I suspected this was the root cause of the dock operator being unable to fully deballast the tank below the level of the pipe.

During the structural inspection, the repair team was cleaning and preparing the two large areas in Tank 10. To maintain the cleanliness of the repair zone, a series of cofferdams were fabricated and fitted into place. Cleaning with fresh water was accomplished by pumping fresh water from the dock ballast tanks into Tank 10. The cracks through the floor facilitated drainage. This is how the rebar sizing and arrangement was determined and how a hole the size of a fist was found in Tank 10, at Frame 88, in way of the WTBHD separating Tank 11. The hole was large enough to see the splintered cribbing below.

To patch this size of crack would require an alternate application method. We had started to prepare the cracks for the Gel Foam following the manufacturer's recommendation. By drilling holes at an angle and intersecting the crack, a shoelace pattern is created that allows for the use of a grease fitting and injecting the foam through the drilled hole into the crack, except when the crack is more than ½". Before cleaning, most of the cracks appeared to be less than that. At the conclusion of cleaning, which included the removal of lose concrete, no crack was less than ½". With cracks larger than ½", the recommendation from the manufacturer was to soak cordage or fabric strips in the gel and packing the cracks.

Much of this information was relayed to the owner via emails sent by the HLV Captain. Our material requests included the Gel Foam 2, 100 gallons of fresh water and bags of clean cotton rags. Upon arrival, our preparations were complete, and the remaining materials were delivered via boat.

The next phase began in earnest as the time to complete them and refloat the dock was 24-hours. Using clean rags, we turned them into 2-inch strips and soaked each one in a bucket of gel just prior to application. Working on each crack in succession, they were packed with gel soaked rags. At this point we abandoned use of the injection fittings for the more arcane method of pouring the gel directly into the crack using the soaked rags as a stopper. Fresh water was then poured from a bucket into the crack which ensured us of a complete reaction of the gel through the crack, most of which we could not see. Within minutes of our first application, we noted that the further addition of water remained in the compartment. Within hours we had completed our patch repairs of all the cracks. Or so we thought.

The following morning, refloating of the dock began. Discussions with the HLV Captain and Load Master led us to the following course of action. All lids to the tanks would remain open during the submergence. The HLV would stop at with 1 meter of water on the hull for watertight checks. If there was a need to abort, the dock would still be firmly on the cribbing and would allow us enough head pressure on the hull to determine if we had successfully sealed the cracks. With a satisfactory watertight check, the HLV would continue submergence until float off and maximum separation, during which time all lids would be sealed for the final tow into port.

As submergence began, nervous anticipation crept in. At 0.5 meters we began to notice that both tank 10 and 11 were filling with water. With 1.0 meters of water the submergence stopped. The Captain and Load Master were notified that we would need time to find the leaks before pumping back up. With the tanks partially submerged, inspection of the repaired cracks found that none were leaking, indicating that there were other unfound leaks. Inspection of Tank 10 revealed a massive crack running transversely at the end watertight bulkhead Frame 68 just behind the repair made to the shear crack at Frame 70. This crack was hidden by a previously made shoddy repair. It appeared that a transverse crack had been previously patched by simply hand packing grout along the edge. This may have held previously, but the damage sustained and subsequent flexure of the dock must have loosed the grout exposing the crack. The appearance after cleaning was that of an unbounded fillet weld, where the weld material can be easily removed as it never bonded to the adjoining material. The unseen crack was large enough to vent air and water through the crack as we were submerging. We attempted to apply Gel Foam 2 directly to the crack, but the unseen portion was so large that the gel poured straight through. Another small crack in Tank 10 at Frame 88 was found near centerline in way of a previously noted surface spall. Here, water was welling up as we submerged. In Tank 11, the water level was slowly rising. Inspection of the tank found a series of cracks in way of the non-watertight longitudinal bulkhead in way of the inboard wing wall and Frame 68 and Frame 70. At this location, just below the pontoon, was an open pipe that that extends from the pontoon deck to the pontoon bottom, presumably for drainage. This pipe had created a hard point when it landed on the cribbing causing the bottom to fracture at multiple points. With the mud cleared away, multiple fractures were found in the bottom shell extending from one side of the non-tight bulkhead to the other as it ran through the drainage hole at the base. No other leaks were found.

Repairs to the new cracks were discussed while the HLV was deballasted. The small hole in Tank 10 near centerline could wait; The multiple fractures in Tank 11 was too large to be repaired quickly; The transverse crack in Tank 10 at Frame 68 could be temporarily patched using Interzone 101, a two-part epoxy. Reiterating the analysis done previously using GHS, it was agreed that the leakage rate had been slowed enough and that the dock, while leaking would not have its stability compromised and we would be able to make port before the level of water in the dock caused a noticeable change in attitude.

With the two-part epoxy patch at Frame 68 complete, final submergence began until float-off and final separation. Plans were already being made for the final repairs, which would start with a 100% dive survey of the bottom with video and surface communication.

The day after arrival, Engineers from Pallada had arrived with the hope they would have more information regarding the construction of the dock. As we watched the divers via video, the dive survey proved to be an engineers version of a Stephen King thriller, as the cracks on the underside were far larger and fractured than what was observed internally. The sizes were such that the use of a cofferdam would not be feasible. Instead the exterior would be sealed with Interzone and internally a double bottom would need to be poured.

The engineers did not have specific information regarding the construction of the dock, nor was there any further construction data found on-board. Their knowledge of the dock was based on their current knowledge of concrete pontoon construction and proved to be quite valuable. [As a point of note, the Russian construction drawings are considered proprietary and would never have been made available. In western countries, we consider that the owner has purchased the design and would never willingly release that information without adequate compensation.]

Wally Sayed
Heger Dry Dock, Inc.

Preparations were also underway to begin thoroughly cleaning the areas which would be double bottomed. The owner had secured a local concrete engineer with experience making concrete repairs to marine structures. Coordination between Heger, Pallada and the concrete engineer developed the final repair plans.

1) Bottom/Underwater repairs. To stop the leaks, divers would apply Interzone externally until all internal leaks ceased. In addition, the concrete engineer would develop a concrete mix, which would adhere to Interzone and to be applied under water which would incorporate carbon fiber sheets to ensure the thin concrete layer would be able to flex without cracking.

2) Interior repairs. A steel reinforced concrete double bottom would be poured in way of all bottom cracks, regardless if they were patched or not. Using information from Pallada, a reinforcement detail was generated which included drilling and anchoring rebar to the dock floor. The intent was to fillpour a double bottom across each affected bay to ensure the longevity of the repairs made. Additionally, reinforcement details were generated in order to repair the bulkheads in way of the lightening hole. The intent was to pour the floor first, then fit a mold around the reinforcement and pour concrete to create a thickened bulkhead in way of the opening.

Prior to my departure, a set of photographs were taken of the reinforcement as it was being installed and fitted.

As a final note, and as a thought question, a very peculiar observance was made regarding the cribbing. After arrival and discussions with the Marine Surveyor regarding the cribbing arrangement, his notes from the docking indicated all the blocks were in full contact at landing. As we inspected the all the cribbing we noticed that even the new cribbing that had been force fitted now had gaps all around. I have a theory, but I'm wondering if anyone has any ideas as to why this might occur?

Industrial & Marine Automation Systems

Floating Dock Control Systems

http://www.dockcontrol.com

ADECS^3

Automated Dock Evolution Control System

MDM

Multi-Axis Deflection Monitoring

Available Dock Data

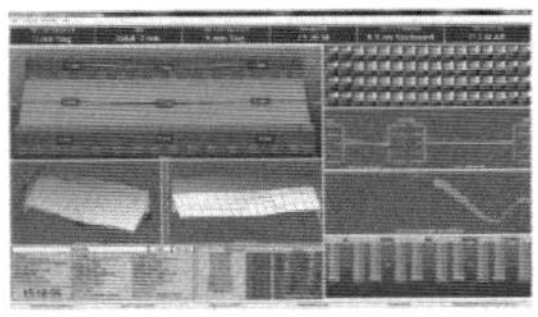

- Trim and List
- Tilt Direction and Angle
- Longitudinal Deflection (Hog/Sag)
- Skew (Twist) and Linearity
- Transverse Deflection (wall in/out),
- Squareness (wall/keel trim)
- Sectional and Wall Specific readings in addition to entire dock readings

Customizable Graphics

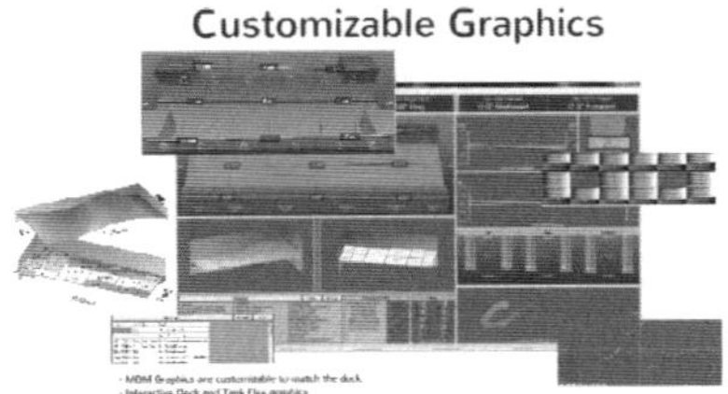

- MDM Graphics are customizable to match the dock.
- Interactive Dock and Tank Flex graphics
- The MDM module works with the rest of the ADECS system to show tank and draft information if needed as well.
- Up to 28 sensors, and multiple master units can be combined to provide redundancy.

How it works

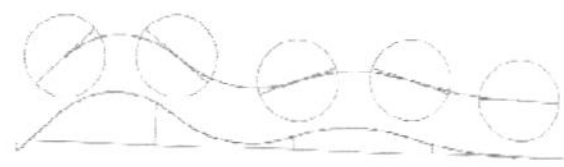

- Each high precision sensor unit reports the state of that particular section of the dock.
- Central processing computer (CPC) reads all sensors and then combines the readings to map out the entire dock.
- An additional sensor is mounted on the keel line to provide a reference point to base transverse bending.

Sensors

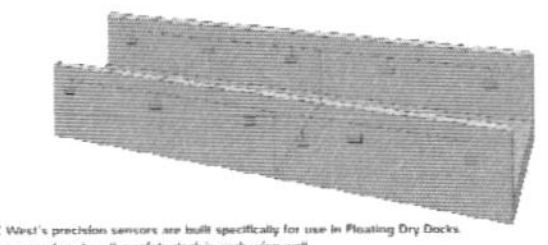

- IDAC West's precision sensors are built specifically for use in Floating Dry Docks.
- Sensors are placed on the safety deck in each wing wall.
- Interconnected by dedicated Power Over Ethernet (POE) Network.
- Central processing computer (CPC) polls sensors, distributes data, and provides graphical display of all values.
- Several Distribution units are located throughout the wing walls to power the sensors and communicate back to the CPC.

RDM

Remote Direct Manual

Remote Manual Control and Monitor

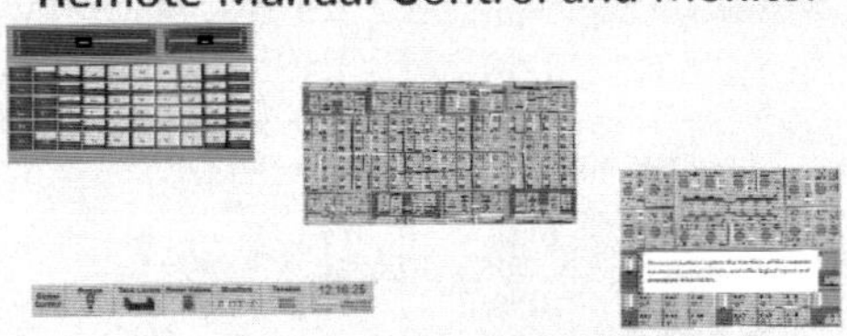

- Touch Screen System for Manual Control, replaces bulky and expensive manual control console
- Redundant – commonly dual panels are supplied, each capable of controlling the entire dock.
- Screens are arranged for operational efficiency based on the customer's method of operation.

Operational details

- Provides display and control functions together in intelligent groupings.
- Keyed switches also included for touch and operational safety.
- Monitors and controls I/O directly without the requirement for any intelligent controller.
- If automatic operation is installed the rest of the system knows that the operator has control of any specific object and reacts accordingly.

Assisted Group Control

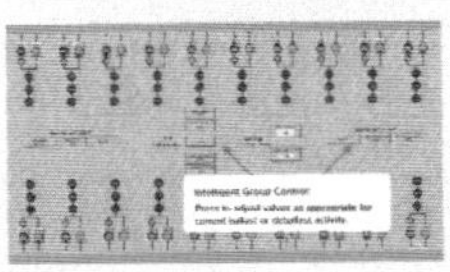
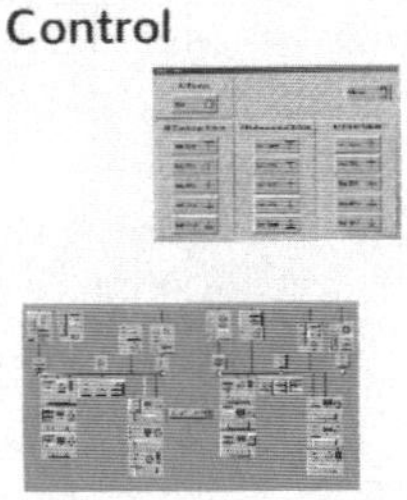

- The RDM module can be used with intelligent group control of valves.
- For instance: ballast and deballast valves could be controlled based on the state of inlet valves or pumps, simplifying list and trim control options.

RAX: Remote Auxiliary

(if desired)

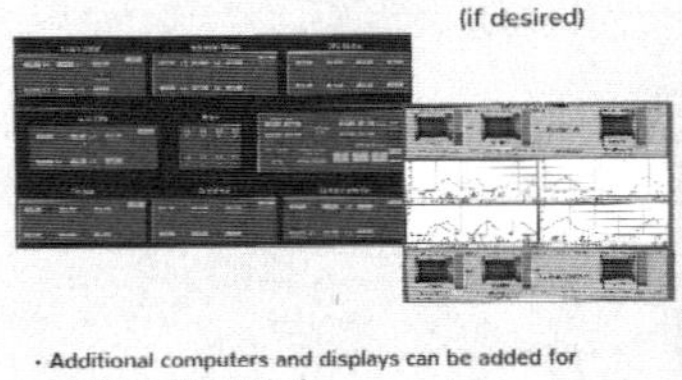
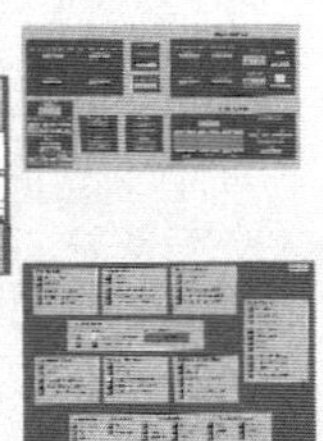

- Additional computers and displays can be added for auxiliary system monitoring.
- The RAX units are also used as additional RDM redundancy.

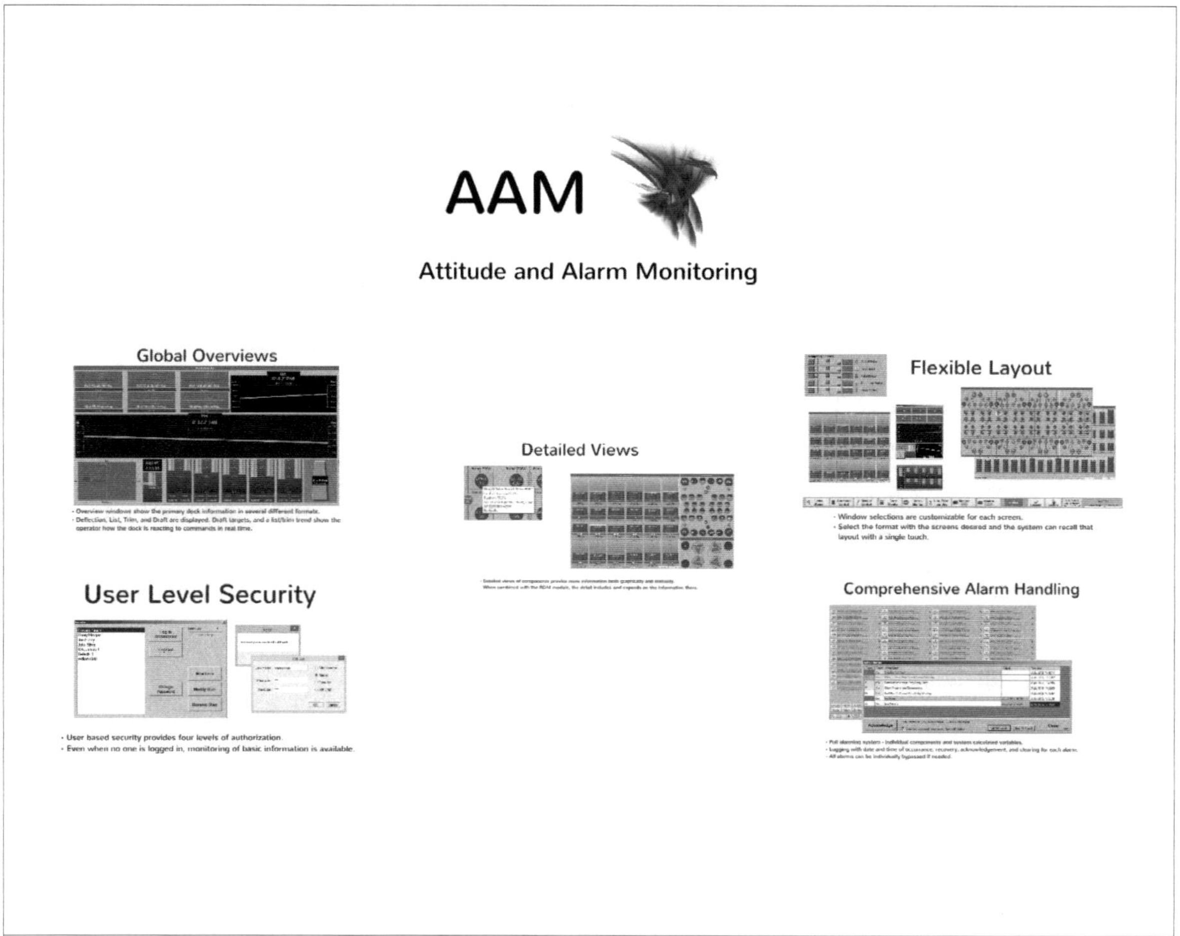
AAM
Attitude and Alarm Monitoring
Global Overviews
- Deflection, List, Trim, and Draft are displayed. Draft targets, and a list/trim trend show the operator how the dock is reacting to commands in real time.
Detailed Views
Flexible Layout
- Window selections are customizable for each screen.
- Select the format with the screens desired and the system can recall that layout with a single touch.
User Level Security
- User based security provides four levels of authorization.
- Even when no one is logged in, monitoring of basic information is available.
Comprehensive Alarm Handling

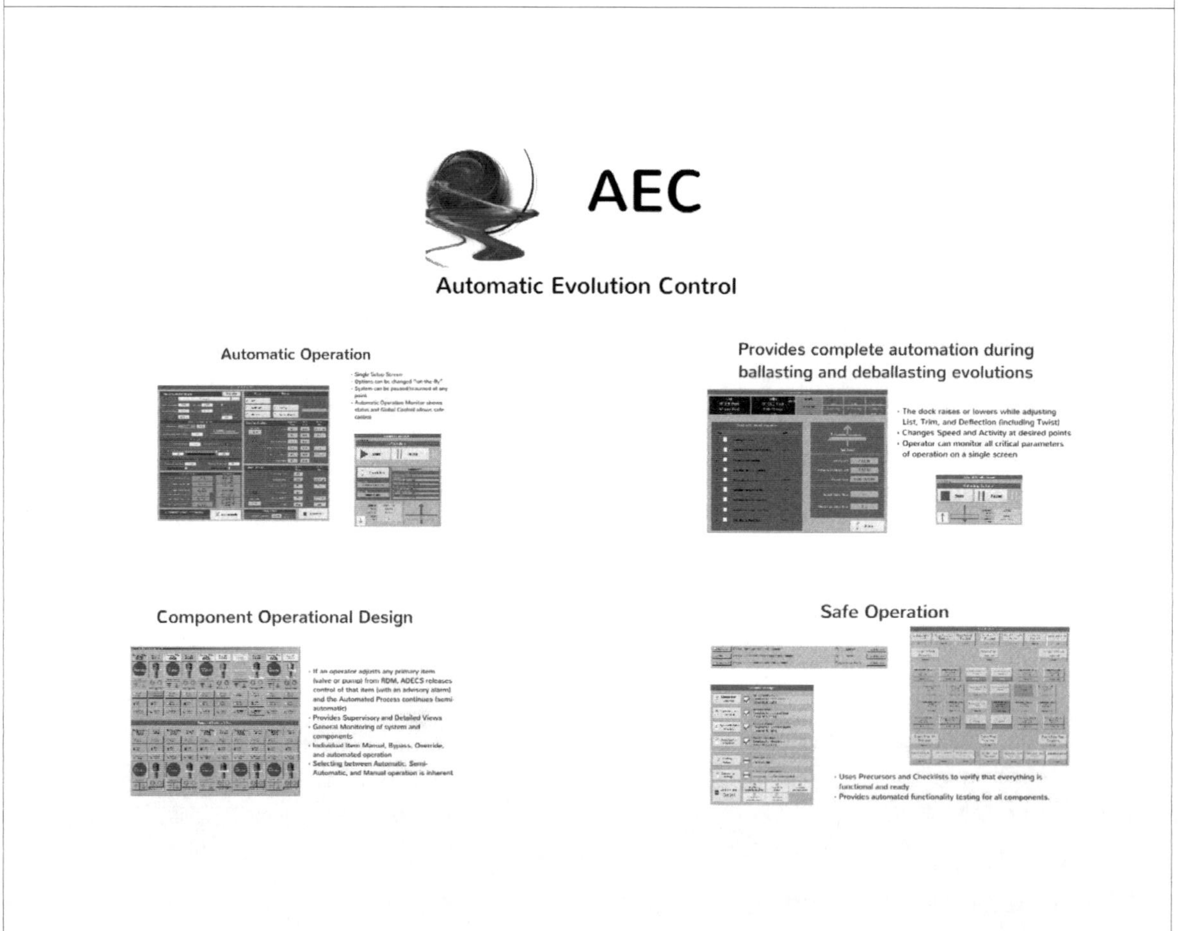
AEC
Automatic Evolution Control
Automatic Operation
Provides complete automation during ballasting and deballasting evolutions
- The dock raises or lowers while adjusting List, Trim, and Deflection (including Twist)
- Changes Speed and Activity at desired points
- Operator can monitor all critical parameters of operation on a single screen
Component Operational Design
- If an operator adjusts any primary item (valve or pump) from RDM, ADECS releases control of that item (with an advisory alarm) and the Automated Process continues (semi-automatic)
- Provides Supervisory and Detailed Views
- General Monitoring of system and components
- Individual Item Manual, Bypass, Override, and automated operation
- Selecting between Automatic, Semi-Automatic, and Manual operation is inherent
Safe Operation
- Uses Precursors and Checklists to verify that everything is functional and ready
- Provides automated functionality testing for all components.

HDL

Historical Trend and Data Log

Real Time Trending

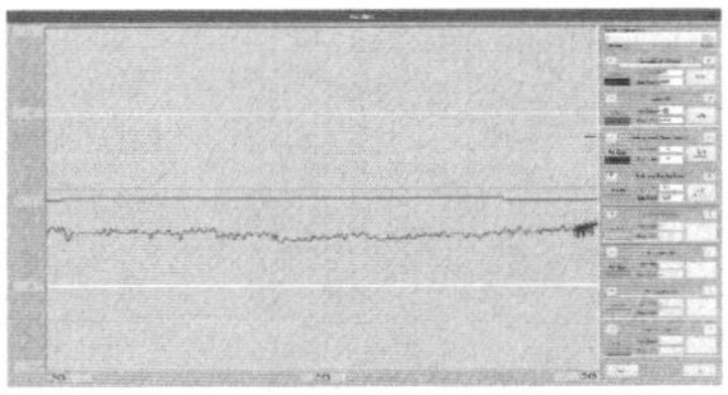

- Live data trending for 5 minutes to 4 hours - on the fly and flexible
- Automatic historic lookup for any item which is historically trended
- Save trend formats and information

Unlimited Event Logging

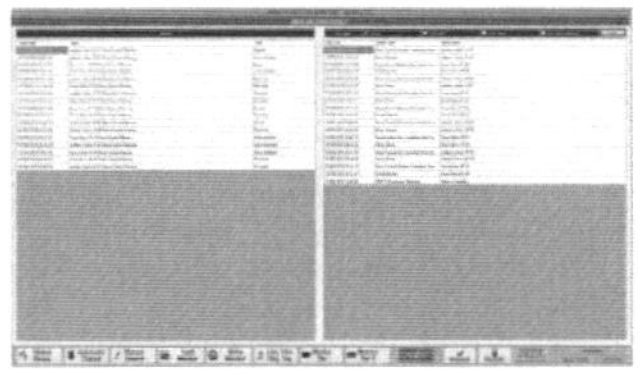

- Events and alarms are logged and stored for an unlimited time
- After two months events are compressed and archived into a public domain format (.zip)
- Events are organized by type and by date/time

History and Historical Export

- No pre-canned report formats
- Select data from history - one minute samples, stored to industry standard CSV files (Excel, Open Office, etc. all read natively)
- Save selected export formats for later ease of re-export
- Events are exported in the same fashion.
- Data is typically stored "live" for 3 months, then automatically archives to a compressed (.zip) file and stored in a separate location
- A standard hard drive will hold about 100 years of information before filling in this manner.
- High Resolution (1 second sampling) historical data, and data export, is also available if required.

Simplified Complete ADECS³ Control System Physical Components

Control Room: Computer Displays

Control Room: CPU Cabinet

Safety Deck: Remote Cabinets & MDM Sensors

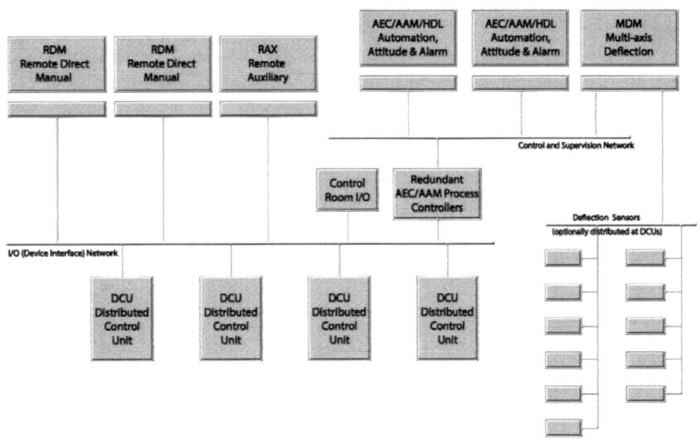

ADECS^3 Control Room Configurations

Designed for:

Operator Capability

The Operator should always be able to control any or all devices – even when the system is running in full automatic
Operator control should include full or partial take-over
Every controlled device should be available for local and remote operator control, bypass, override

Backup / Redundancy

- If any subsystem fails, there should still be a way to safely operate the dock.
- Any critical components should be redundant.
- Critical Information should always have a second source of data.

Hardware — Standards — Software

Hardware

Commercial Off-The-Shelf (COTS) Components

- Opto 22 PAC Controllers and I/O
- Industrial Computers and Touchscreen Displays
- Ethernet Networks with standard hardware
- All replaceable locally – where the dock resides

Specialized Components only where needed

- Specialized Sensors
- Backup parts available on site
- Critical COTS parts are pre-loaded

Software

Industry Standard Core

- Microsoft Windows
- TCP/IP and FTP Communications Protocols
- Opto 22 Flow Charting Language and Graphical User Interface
- Critical software written in Microsoft Visual Studio

Source Code and Development

- Development Software Licenses are included
- Source Code is included in final delivery
- Backups of all source code is kept indefinitely (at least 10 years) in both cloud and local by IDAC West
- All configuration files are stored with the source code - including a complete revision history

ADECS3

Automated Dock Evolution Control System

ADECS3: MDM
Multi-axis Deflection Monitor Module

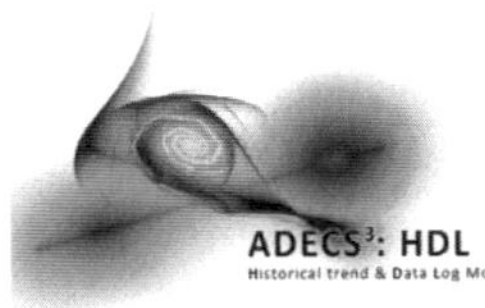

ADECS3: HDL
Historical trend & Data Log Module

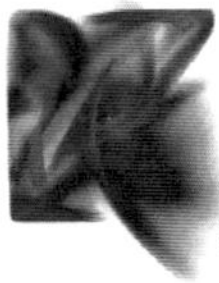

ADECS3: RDM
Remote Direct Manual Control Module

ADECS3: AEC
Automatic Evolution Control Module

ADECS3: AAM
Attitude & Alarm Monitoring Module

About IDAC West

History

- Providing State of the art automation systems since 1993
  - Global Projects - From Silicon Valley to the South Pacific
  - Principle Staff - Each with 30+ years of controls experience
  - We have written many of our customers' standards
- Marine Automation Systems
  - Designing and Maintaining Floating Dry Dock Control Systems since 1999
  - All of our marine customers are still our customers to this day
  - Currently the US Navy standard

Capabilities

- Broad foundation of control and process skills
  - Automation System Experts
  - Wide variety of PAC, computer, database, and web programming skills.
  - Electrical Engineering, Sensor Development, Electronic Control Design
  - Mechanical Engineering Staff
- Industry Expertise
  - Floating Dry Docks
  - Water Distribution, Material Handling
  - Food Processing, Semiconducter
- Additional Strengths
  - CAD and 3D CAD Modeling
  - Panel Manufacturing
  - Project Management

References

- Floating Dry Dock Systems
- US Navy, Point Loma CA
- Australian Marine Complex, Henderson, Australia
- US Navy, Philadelphia PA
- Electric Boat, Groton CT
- NASSCO, San Diego CA
- BAE Shipyard, San Diego CA
- Qatar Petroleum/Hyundai, Ras Laffan, Qatar
- KeppelFels Brasil, Angra Dos Reis, Brazil

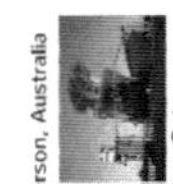

- Classification Societies
  - NAVSEA
  - Bureau Veritas
  - ABS
  - Lloyds Register

Support

Local Support

- Worldwide Local Support
  - Local Hardware Support via Opto 22 Reps
  - IDAC West Support Agreements for software
- Purpose of Local Support
  - First Point of Contact
  - Primary support for hardware issues
  - Basic software configuration support
  - Follow-up Maintenance personnel training
- Additional Functionality
  - General system updates handled locally
  - System Maintenance reviews
  - Spare parts inventory

Expert Support

- Expert level issues are handled by remote connection
  - Secure internet connection
  - Available only when activated by local personnel
  - Allows system developers to access core parameters and programming.
  - All activity is monitored and visible to local personnel
- Time frames
  - During initial year an expert will be online monitoring operation for each scheduled evolution
  - Normal expert response time for issues is 2 to 4 hours after notification
  - Expert support will be on site during start-up and commissioning

Training

- Computer Based Trainer/Simulator
  - Acts the same as the dock
  - Uses same code/views as in the control room
  - Allows operators to see how the dock control system reacts to situations and to learn how best to correct them
- Initial On-site Training
  - Operations Training
  - Maintenance Personnel Training
- Follow-up Training
  - Operations Training using CBT and Self-guided training material
  - Hardware/Maintenance training by Authorized Service partner

An Alternative Dry Dock Ship Supports Method

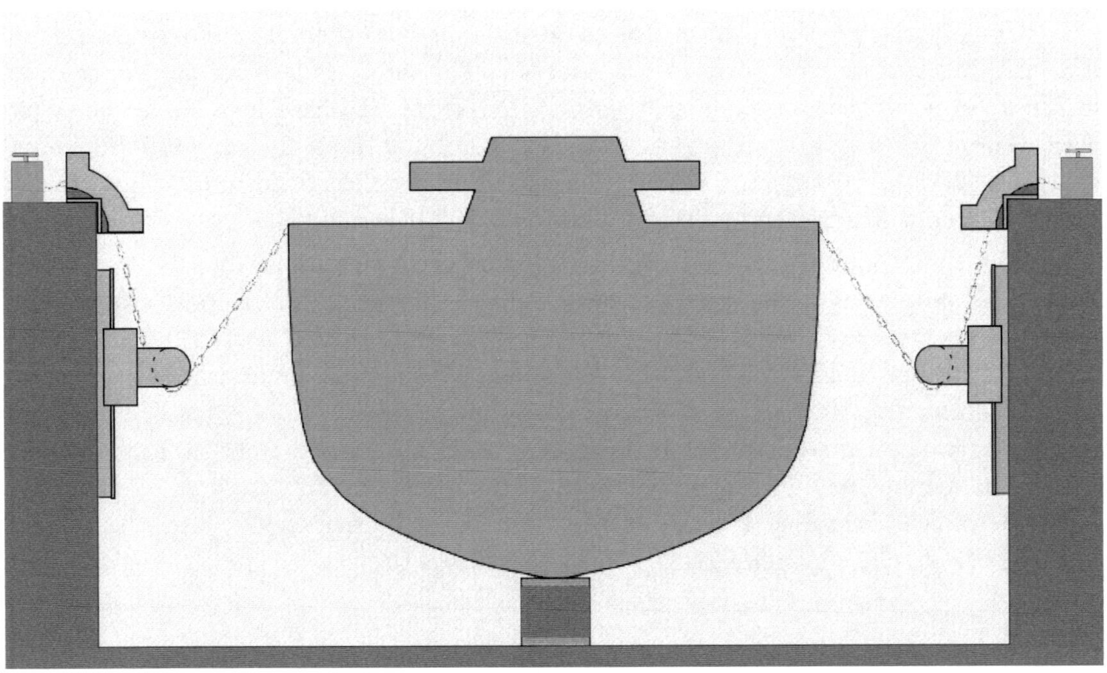

Alexander Stiglich

April 12, 2017

Contributions by:
Jonathan Leyva
Adam Johnson

Sponsored by:
DM Consulting

ASME awarded “Best Project” for MAE 156B Winter Quarter 2017, UCSD

Abstract

The objective of this project was to design a new method of supporting ships in dry dock and support the design with analysis. The new drydocking method was to design a support for the US Navy Frigate (FFG-7) in the San Diego Naval Graving Dock. The FFG-7 is a medium size surface combat ship 135 [m] (450 [ft]) long weighing 3200 [Mt] (3100 long tons [LT]) under light loading conditions. The graving dock dimensions are 200 [m] x 30 [m] (700 [ft] x 100 [ft]).

The most popular current method of supporting ships in dry dock is using blocks. The blocks method consists of centered keel blocks which support most of the weight of a vessel and angled side blocks on each side to hold the ship upright for stability. The side blocks must be cut to the shape of the vessel and be placed precisely to avoid damaging the vessel. Custom fitting each vessel takes a lot of time and resources. The new method developed replaces these side blocks with a chain system. The chain system could reduce the preparation time to dry dock, the resources required, and the overall risk. The chain system was designed based on the standards required of side block systems by US Navy standards (Ref 1).

Table of Contents

Chapter 1: Project Description

1.1 Background

1.1.1 Drydocking Explained

Drydocking is an old industry. Since ships were first constructed, there has been a need to take them out of water to repair them, hence the act of drydocking. The most popular method of drydocking currently is using blocks. The blocks consist of centered keel blocks which hold most of the weight of a vessel and angled side blocks (also known as bilge blocks) on each side to hold the ship upright for stability. See Figure 1.1.1 showing a cross section of a ship on blocks.

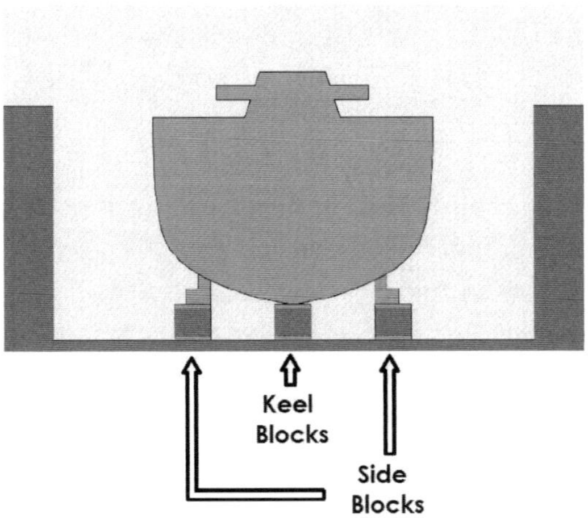

Figure 1.1.1: Cross section graphic identifying keel blocks and side blocks.

A ship will enter the dock full of water, the dock will be sealed, and the water will be drained, allowing the ship to land on the blocks. See Figure 1.1.2 displaying diagrams of cross sections of a ship and a dry dock, illustrating the docking process.

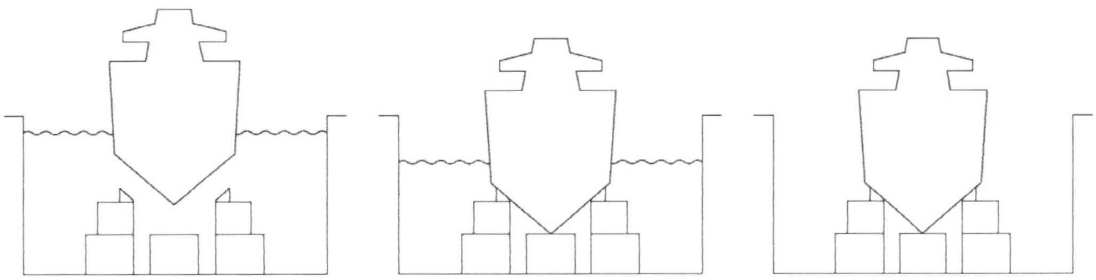

Ship enters the dry dock. The dock is sealed.	Water is pumped out lowering the ship onto the blocks	The dry dock is pumped dry. The ship has been drydocked.

Figure 1.1.2: 3 panel diagram of cross sections of a ship and a dry dock illustrating the docking process

1.1.2 Project Origin

This project was originally a Mechanical Engineering senior design team project for Alex Stiglich and a few of his student peers at University of California: San Diego. The other students involved in the research and analysis of this project were Jonathan Leyva, Adam Johnson, and Yao Chen. Other students of the MAE 156 - Fundamental Principles of Mechanical Design courses also contributed to project. The instructor, Dr. Jerry Tustaniwskyj and teaching assistant, Pedro Franco Navarro who also contributed and supported the project along the way.

The project was sponsored by Joe Stiglich of DM Consulting. Joe is the current President and Chief Engineer of DM Consulting Inc. DM Consulting, headquartered in San Diego, provides dry dock training, software, conferences, and consulting for organizations around the world. Joe Stiglich is a retired US Navy Docking Officer with a Master's of Science in Naval Architecture and Marine Engineering from Massachusetts Institute of Technology (MIT). Responsible for over 300 safe docking operations, he has an insightful perspective to the industry's successes as well as areas that could use improvement. As the sponsor of this project, he was responsible for supervising and critiquing the progressive design development the project. The dry dock industry has been slow in improving its methodology, DM Consulting is dedicated to introducing a proficient, cost-efficient, safer system to the drydocking industry.

1.2 Review of Existing Ship Support Systems

1.2.1 Side Blocks

The most prominent system of drydocking currently consists of using centered keel blocks which support most of the weight of a vessel and angled side blocks on each side of the vessel which primarily keep the ship upright and stable. See Figure 1.2.1.1 which identifies side blocks and keel blocks supporting a ship. The side blocks are cut to the shape of the vessel and are placed precisely to avoid damaging the vessel. The top surface of the side blocks ideally should fit evenly against the ship's hull. Custom fitting side blocks for each vessel requires both a great deal of time and resources. The keel blocks often do not need to be cut to shape so they do not require much time or resources and therefore will remain unchanged this project. The chain system only replaces the side blocks.

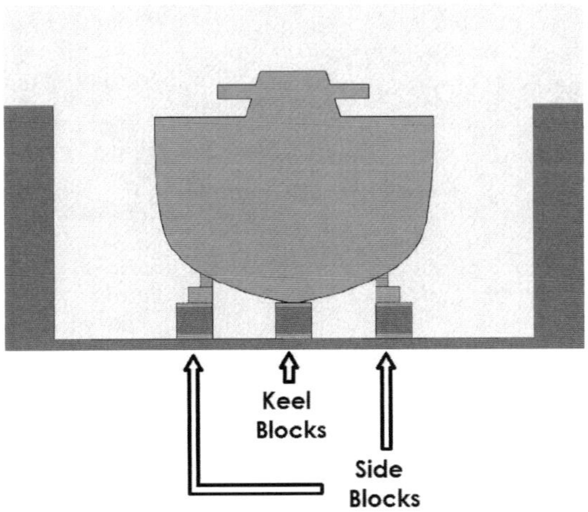

Figure 1.2.1.1: Cross section graphic identifying keel blocks and side blocks.

The side/keel block system is successful in keeping a ship stable for the drydocking process, however, there are a few concerns raised by this process. First, the specific curvature of the ship's hull is very import to match accurately. When a ship is in the water, the pressure the water has on the hull of the ship is distributed evenly. Out of the water, the block supports load the hull at structural strength points. The contact area of the blocks are minimized to allow for access to the hull and reduce the side block production cost. Shaping the blocks to the shape of the ship is also difficult and time consuming for ships that are not just flat bottom, like the FFG 7. These type of ships require tailoring the blocks to specific shapes. This method requires more cost of cutting the side blocks to shape and adds more risk of damage to the vessel.

The materials used in side blocks are exposed to seawater, environmental conditions, and compressed repeatedly. These factors limit the life span of the materials. The blocks are primarily composed of softwood, hardwood, and concrete. Concrete has a longer life than the wood materials. The softwood needs to be replaced often. This is economically expensive and not ecologically friendly.

1.2.2 Shores

Shores are supports extended from the walls of the dry dock to the sides of the ship. They are typically beams or posts made from timber used to support the vessel in lieu of side blocks. See Figure 1.2.2.1 for a diagram of a cross section of a ship drydocked with shores. The primary problem with this design is in the case of an earthquake the ship may shift slightly possibly resulting in the shores falling into the basin. This would cause ship to fall endangering the lives of any workers during maintenance operation.

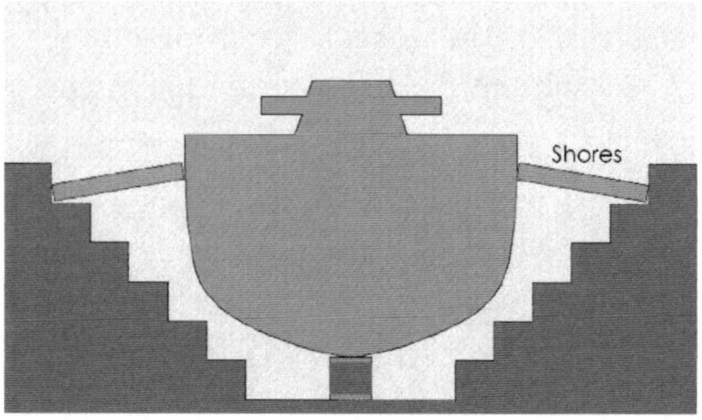

Figure 1.2.2.1: Diagram illustrating cross sections of a ship in dry dock supported with shores

There are issues with the installation of the shores as well. They must be installed synchronously and can be heavy and difficult to install. Their geometrical length requires adjustments for different ships. Shores must have some support in the transverse direction on the dock. Older dry docks were shaped with steps to allow for the shores to be installed at different heights and still have the transverse support. The advantages of shores are, by nature they are not directly exposed to seawater and completely visible during installation.

1.2.3 Chains

The use of chains to support a ship in dry docking has already been implemented at a shipyard in the Port of Rotterdam, South Holland in the Netherlands. See Figure 1.2.3.1 a diagram of a chain system. Like shores, chains allow for full visibility of the stabilizing system during installation. The use of chains would also eliminate the concern of seawater exposure of the drydocking system. Before the dock is drained, one end of the chain is attached to a buoy and floated across to the ship. This process would also save on monetary costs, since the side blocks would no longer be deemed necessary.

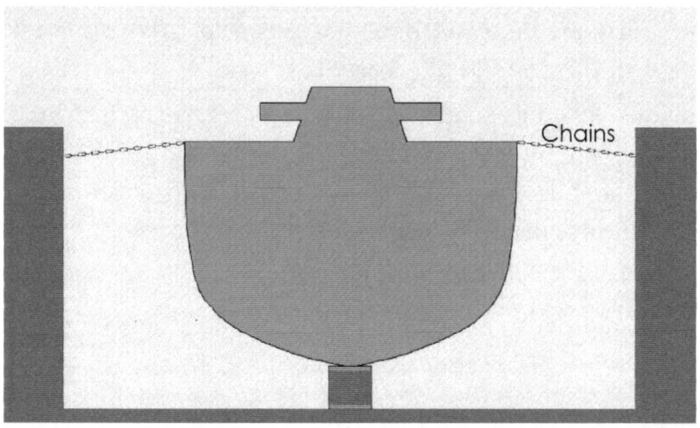

There are many questions as to how this system got started and why it is so uncommon. The Netherlands shipyard was contacted regarding the engineering behind their method of drydocking. They responded to some of the design questions that were posed, however, they did not have engineering analysis behind their system. The current yard operators do not know neither the engineering nor the history of the chain system, only how to use it. They are getting excellent results with short operational periods. With an existing operational system in use, it was decided that despite its limited use in drydocking, the chain system is a functioning mechanism that warranted additional study. The basic idea applied with proper engineering analysis has the potential to be a revolutionary method of supporting ships in dry docks.

1.3 Project Objectives

The purpose of this project is to explore and demonstrate an alternative dry dock ship support method. The drydocking industry has been slow to make changes and improve over the years, this project hopes to push the field towards innovation. Side blocks, the current dry dock method that most ship-yards use is outdated, wasteful, and inefficient. A better system could conceivably make dry docking safer, faster, and less expensive.

This project specified to develop a more efficient way to dry dock a US Navy Frigate (FFG-7) in the San Diego Naval Graving Dock. See Figures 1.3.1, 1.3.2.

Figure 1.3.1: FFG-7

Figure 1.3.2: San Diego Naval Graving Dock

The FFG-7 is a medium size surface combat ship 135 [m] (450 [ft]) weighing 3200 [Mt] (3100 [LT]) under light loading conditions (drydocking conditions). The San Diego Naval Graving Dock is a cement dry dock 200 [m] (700 [ft]) long, 30[m] (100 [ft]) wide, and 12 [m] (40) [ft]) deep.

Objectives:

- Design a new method of dry dock ship supports replacing side blocks
- The system is made for drydocking the FFG-7 in the San Diego Naval Graving Dock
- The design should have improved cost, efficiency, and safety over current side block method
- Improve and justify the design with engineering analysis
- The design must comply with dry dock industry standards (Ref 1)

1.4 Functional Requirements

Transverse Stability

From a dynamic perspective, the current system acts as an inverse pendulum where most of the weight of the ship rests on the keel blocks and the side blocks provide transverse support. The system must resist against an external force acting at the center of mass which applies a moment overturning the ship.

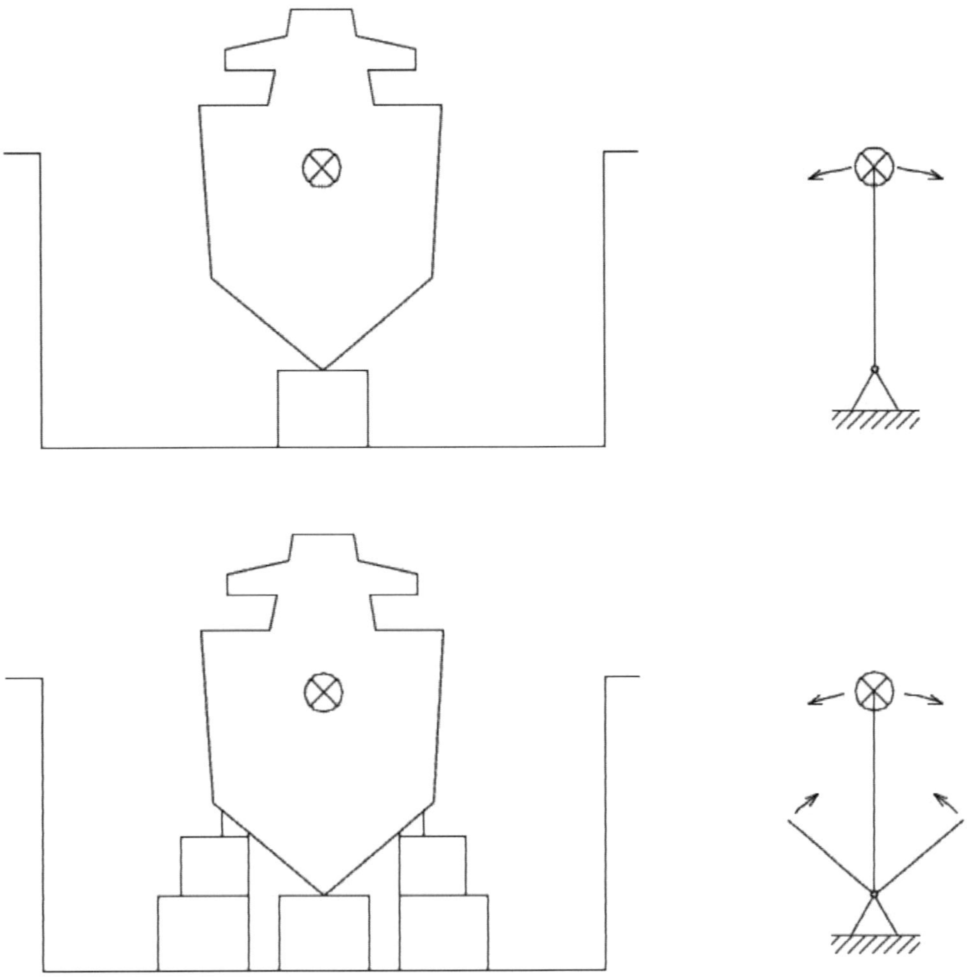

Figure 1.4.1: Inverse pendulum of a cross section of a ship pivoting on the keel blocks

The system is not for drydocking ships without any shape, as seen in Figure 1.4.2. These ships are docked with flat side blocks with little risk of falling off the blocks transversely.

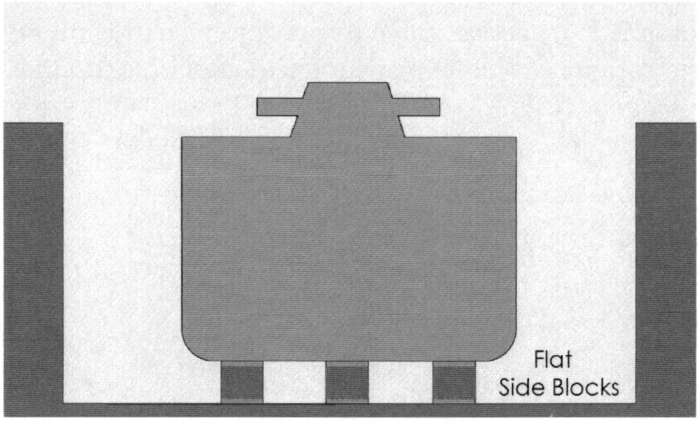

Figure 1.4.2: Diagram of cross section of a ship without shape drydocked with flat side blocks.

Primary Objectives

Economic cost and time efficiency:

The new design must cut costs. The lifetime monetary expenses should be reduced from the current values. This includes the pricing of replacing materials as well as the costs for the workers. Additionally, less workers results in lower costs. Aside from the financial costs, the time should be reduced as well. The time saved during the drydocking process should be reduced creating a more efficient process, another appealing factor to the market.

System Failure Prevention

The new system ideally should not rely on any single component. Multiple component failure would be required for a whole system failure.

Loading

The system should withstand emergent loads, such as those caused by earthquakes and hurricanes. For the FFG-7 under a standard earthquake (Ref 1) provides a given load of approximately 4,982 kN (500 LT).

Stability

The system should be stable in both the transverse and longitudinal directions.

Demonstration

Scaled models of different components as well as the entire system should display the new proposed system. A 1:350 scale model of the system provides a view of the system with the ship in dry dock. Other scaled components demonstrate the engineering analysis by showing stability in a simulated earthquake.

Secondary Objectives

Adaptation

The new system must be designed in such a way that it can be implemented in other dry docks. It must also be adaptable to various types and sizes of ships. The wider the range of possible ships and dry docks, the greater the chance of successful implementation in the industry.

Application

The system should be adaptable to existing dry docks. The design fit into existing dry docks with the least amount of modifications to the existing structure.

Constraints

Salt Water Exposure

One of the problems with the current system of dry docking is the exposure of all the components to sea water and the corresponding corrosion. The financial costs of constantly replacing materials such as softwood accumulate. Ideally, the system should operate entirely out of the water.

Maintenance Clearance

The system should leave sufficient open space in the dry dock so that ship maintenance operations can be performed safely and efficiently. There should be enough room to allow for passage of a fork truck around the ship. There should also be enough room for easy access to the ship's hull for preservation and repairs.

Chapter 2: Final Design Solution

2.1 Final Design Solution Summary

The selected chains method of dry docking was progressively developed and enhanced over the course of the project. The original idea was to secure chains to various points along the sides of the ship. See Figure 2.1.1.

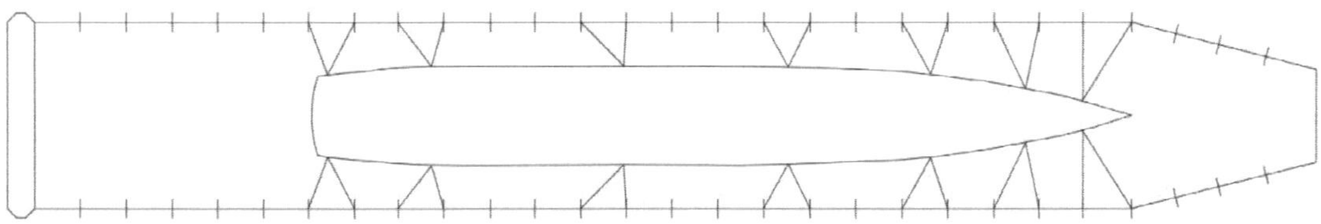

Figure 2.1.1: Diagrams of the ship in dock with chain placement

The chains are secured to the bitts of the ship. The chains are anchored to the wall with multiple components: the rail, the tension system, the static upper chock, and the mobile lower chock. A profile view diagram of the rail system on the concrete wall can be seen in Figure 2.1.3. Rather than simply securing the chains to a fixed

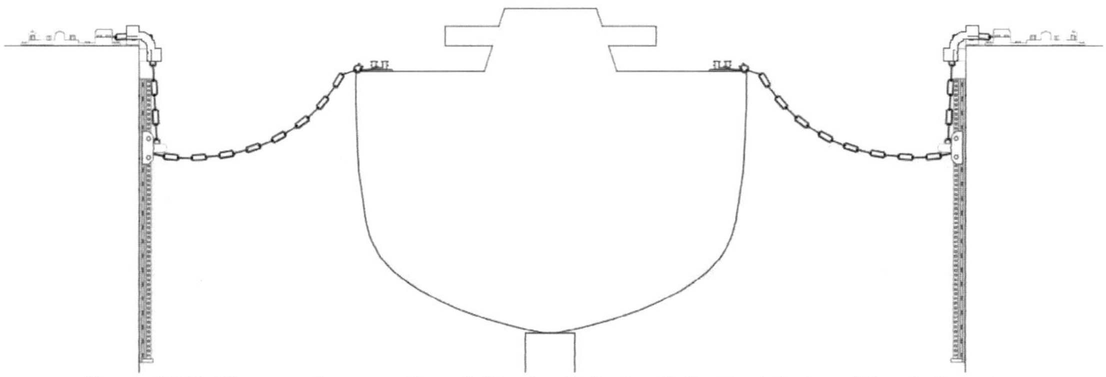

Figure 2.1.2: Diagram of cross section of ship dry docked with the final design of the chain system

point, this system is designed so that the angle and tension of the chain can be adjusted. The chain is secured using a tension system located on top of the cement wall. The chain is then fed through the upper corner chock component and down along the rail through the lower chock along the wall. From here, then chain extends to the bitts on the ship. The lower chock is mobile in the vertical direction. This allows the angle of the chain to be adjusted, which ultimately results in an optimized tension in the chain. Each of these rail systems is spaced about 7.62 m (25 ft) apart longitudinally from each other along the dock. While there can be numerous rail systems along the wall, they do not take up much space, so there is still much room for standard operations during the lay period.

There is still more analysis that can be completed, and testing that needs to be done before a ship should be drydocked with this method. If it is determined there is excessive movement in the chain, a spring-damper could be installed. The tension system component could be replaced with a spring-dampener mechanism. A profile view can be seen in Figure 2.1.4. This is designed to allow the appropriate adjustment of the tension levels in the chain. If this were implemented, it would replace every other tension system component longitudinally (every 15.24 [m] (50 [ft])) along the dock. The chains would be passed through the connection on the ship to allow for one continuous chain between two dock side attachments so one spring-damper system is acting on each chain.

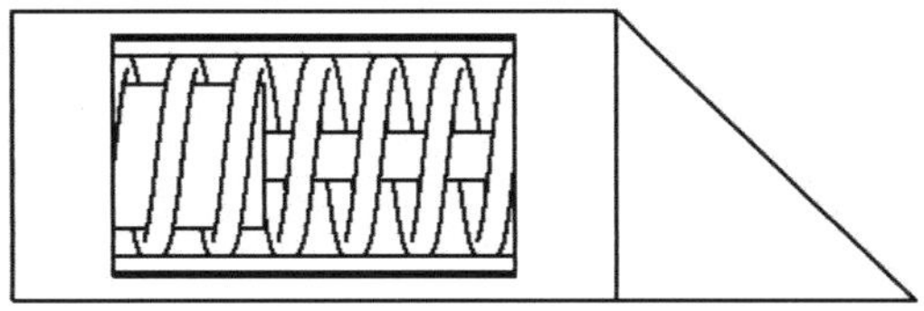

Figure 2.1.4: Profile View of Spring-Damper Optional Component

The length of rail can be adjusted to various heights according to the expected ship heights. Other options where the rail can be used on the top of the dock may be applicable too. Heavier ships will require larger chains and decrease system accessibility and increase the monetary expenses.

2.2 Operational Process

Preparation

Before the ship enters the dry dock, the dock needs to be prepared for operation. The ship information, including the dimensional specifications, locations of bitts and chocks, size/ratings of bitts and chocks, and all other normal dry dock related information (hydrostatic data, ship loading, etc), should be gathered prior to the ship's arrival. Dry dock calculations are completed. The lower chocks are vertically adjusted with the chains with cranes on the rails. The keel blocks are adjusted for the ship.

Drydocking Process

The dry dock if flooded. The ship is hauled into the dry dock using tugs and line handlers. Once it is approximately in position above the keel blocks, the caisson is closed (the caisson is the gate that seals off the dry dock), and the chains are hauled to the ship via crane or floated over on buoys and lifted to the ship by hooks. The line handling lines are cast off. After the dry dock is sealed off, the water de-watering process is initiated and the chains can be used to center the ship. The ship lands on the keel blocks and the dock continues to de-water. After 0.3 [m] (1 [ft]) of water is taken of the hull, the chains tensioned to the specified holding tension. It is important to have the chains at the holding tension before the ship draft is at the draft at instability (Ref 1). It is important that chains on each side are pulled into tension synchronously. This does not mean that all chains in the entire system need to be tensioned simultaneously, but the two chains attached transversely to opposite bitts of the ship must be tensioned simultaneously. This will avoid inducing a list in the vessel, since pulling one side before the other would cause an imbalance in applied forces. The chains are locked into place and the rest of the water is drained from the dry dock.

Undocking Process

Ship flood and leak checks are conducted prior to floating the ship. The dry dock is flooded until the ship has floated 0.3 [m] (1 [ft]) above the keel blocks. The chains are relieved of their holding tension. The opposing chains are relieved of tension synchronously, analogous to the drydocking procedure. The dock is flooded to the tide level. The caisson is then removed and the chains are detached and replaced with line handling lines. The chains are hauled off the ship to the dock, analogous to the drydocking procedure. Tugs then remove the ship from the dock with the help of line handlers.

2.3 Scale Models

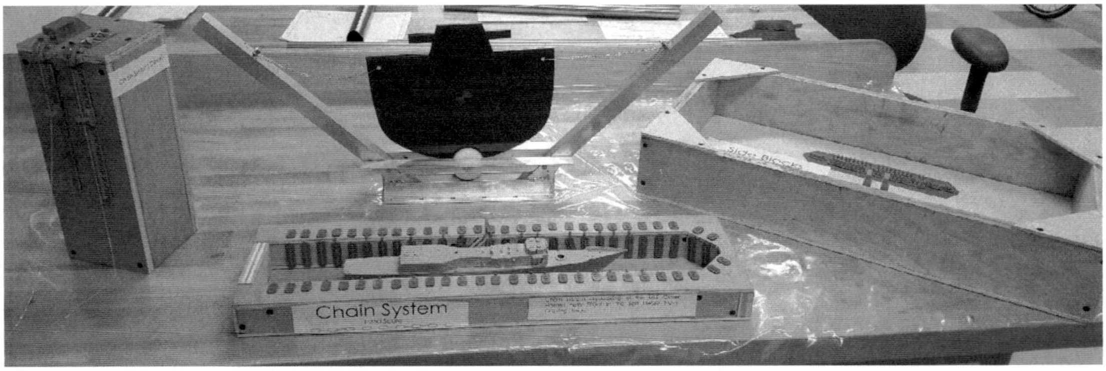

Figure 2.3.1: Photograph depicting all scale models developed for this project

2.3.1 1:350 Scale System

See Figures 2.3.1.1 and 2.3.1.2. A 1:350 scale model of the while system with the FFG-7 in dock was built. This is for visual purposes to provide an example of how the system would ultimately appear. This included a scaled model of the ship inside a scaled dry dock drydocked with the proposed chain system. A miniature 1:350 scale replica of the FFG-7 ship was used. The scaled dry dock was constructed using plywood. Two parts of the dry dock were removable, the caisson, and a section of the dry dock wall, which slide on and off of the pins vertically. These sections were removable to allow viewers to see inside the dry dock and view the keel blocks. The keel blocks were designed to scale in AutoCAD Inventor from the FFG-7 Docking Plan (Ref 2). It was 3D-printed on a MakerBot in PLA. The rail systems were also designed in Inventor and 3D-printed as separate individual components. These prints included the rail with the lower chock attached, the upper chock, and the tension system. Metal-coated thread was used to portray the chains in the system. Each one was attached to individual bitts on the ship replica.

A larger scale model was also created to provide a closer look at the chain system by displaying individual rails. The pieces were 3D-printed 1:33 scale. See Figure 2.3.1.3. The model demonstrates both the primary tension system as well as the secondary spring-dampener system that could be implemented. If the primary chain system proved to be insufficient, the spring-dampener system would be implemented. This is essentially the same design as the primary chain system, but the tension system component on top of the wall would be replaced with a spring-dampener component. A small chain was attached to the tension system and fed through the two chocks. This effectively gave a visual of the chain system.

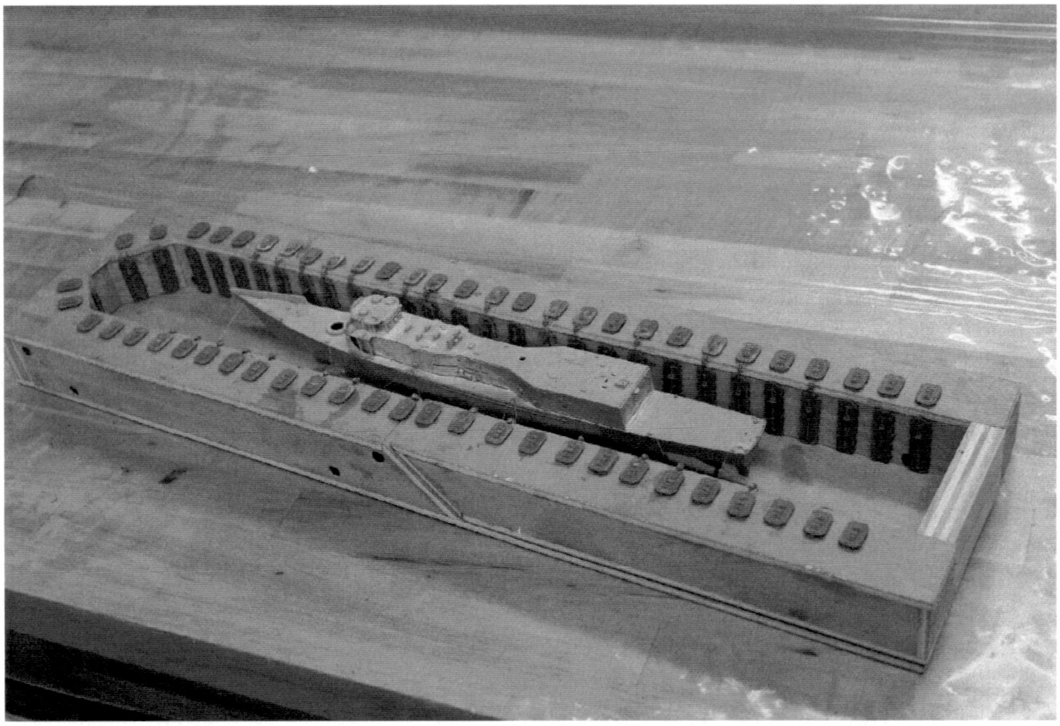

Figure 2.3.1.1: Photograph depicting the chain system scale model (1:350 scale)

Figure 2.3.1.2: Photograph depicting the chain system scale model (1:350 scale)

Figure 2.3.3.1: Photograph depicting the 3D-printed chain system model detail display

2.3.2 1:350 Side Block Display

See Figures 2.3.2.1 and 2.3.2.2. The side blocks, just like the keel blocks were in Inventor to the Docking Plan spec (Ref 2) and 3D-printed. The side blocks were included to provide a visual of the current existing dry docking system. The replica model rested on the blocks quite smoothly, demonstrating the accuracy in the modeled print.

A second 3D-print was made that displayed a small section of the block system but at a larger scale (1:33). With this model, one can more clearly see the arrangement of the keel and side blocks relative to each other.

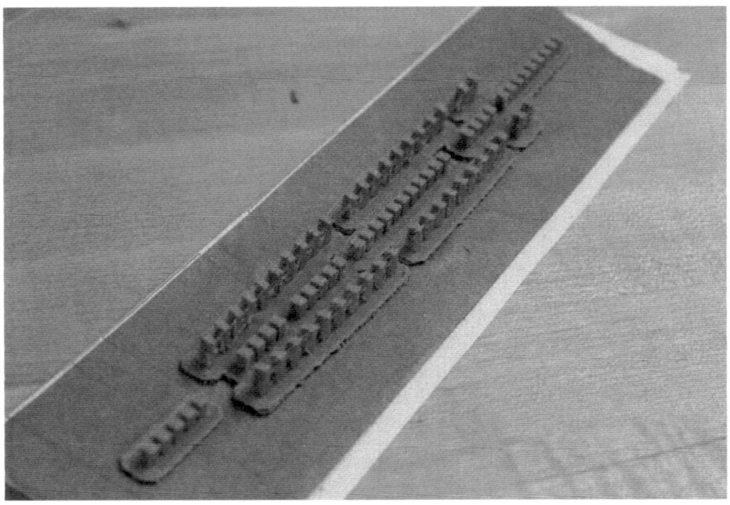
Figure 2.3.2.1: Photograph depicting the side blocks 3D-printed model (1:350 scale)

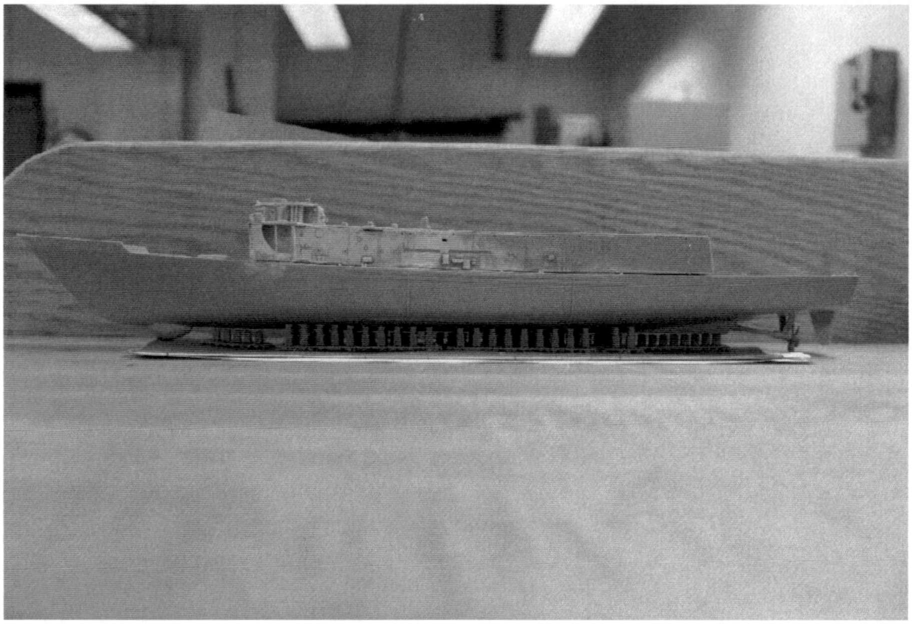

Figure 2.3.2.2: Photograph depicting the side blocks 3D-printed model drydocking the model FFG-7 (1:350 scale)

2.3.4 Earthquake Simulation

This model was designed for more than just visual appeal. It successfully demonstrates vibrational analysis of the system by simulating a cross-section of the ship during a scaled earthquake. Using a shaker table, the system would shake side to side with given input frequencies.

A cross-section of the ship was modeled using AutoCAD based on a cross section of the ship around the mid-perpendicular from the Docking Plan (Ref 2). It was cut out of acrylic using an industrial laser cutter. A frame was cut from 1 [in] square tubing and welded together. The small-scaled chains were attached from the acrylic cross-section to the metal sides. Delrin spacers were cut on a lathe and placed on both sides of the cross-section. A pin extended through the frame, spacers, and cross-section in the middle so that rotation could occur about that point. It allowed the cross-section to tilt in the transverse directions.

A shaker table was utilized from the structural engineering department at UC San Diego. The model was tightly screwed onto the platform. A software program associated with the NEES Instructional Shake Table (ITS100) was downloaded, and the shaker table was connected to both the computer and a controller using the appropriate wiring. The NEES database contains a series of historically recorded earthquakes, which are scaled and saved as simulation files. The final step was to run the desired earthquake simulations, which shook the table and model for a short period of time. From this, the vibrational stability of the system during hazardous conditions could be examined.

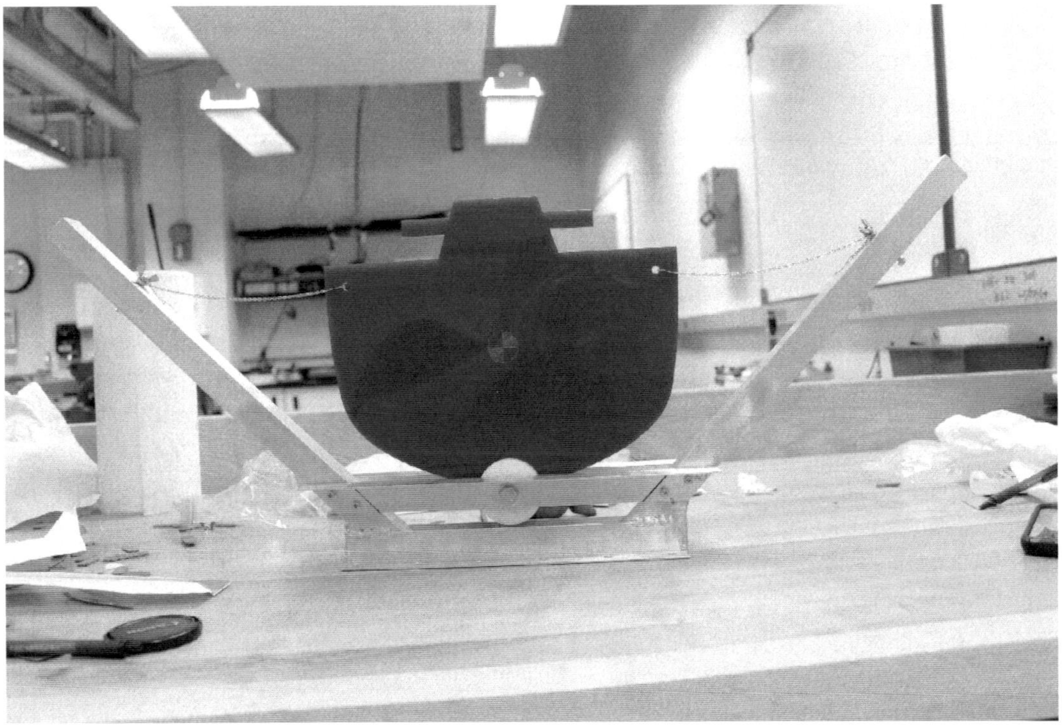

Figure 2.3.4.1: Photograph depicting the Vibrations Test Rig with mounted ship cross-section

Theoretical Predictions

The test model was tested on a shaker table designed to apply specific earthquake loading for the system. This was applied by a spectrum-analyzer/waveform generator to a pneumatic oscillator.

The non-linearity of the catenary equation should not be excited under cyclic loading. The resonance frequency of a spring mass system is a function of the spring constant and the mass. However, if the spring constant is not a constant, there should not be a specific resonance frequency. Therefore, the non-linearity does not allow a system to become excited under simulated scaled earthquake loading.

By tightening the chains of the vibrations test model, the jerking force of the ship on the chains is reduced.

Tests and Conditions

A vibrations test was performed on the 1:50 scaled model by subjecting it to an earthquake simulation. The model consisted of a scaled cross section of the ship weighing 332.5 grams, the frame representing the dock and a set of chains weighing 0.0993 grams/cm to represent our docking system. See Figure 4.1. The model was pinned at the point where the ship meets the keel to limit the variables of the test. This eliminated testing slip from this experiment, however this could be easily addressed with addition of rubber blocks. The testing of this model was done using a shake table, its operating interface and an Arduino. The model was subjected to three different earthquakes varying in magnitude. The smallest simulation was a 0.38 scaled version of the Kobe earthquake that happened in Japan on January 17,1995. The other two simulations were of earthquakes stronger than the Kobe earthquake but corresponded to no real earthquakes.

Results

From the various vibrational tests, it was determined that there is a specific chain tension for any given vessel that will prevent ship movement and chain nodal oscillation. As was evident from the catenary analysis, this point is approached as the slack in the chains approaches zero. However, due to the real chains' diameter and weight, this tautness cannot be reached. In addition, trying to apply these results would put the ship under constant unnecessary tension. After reducing the tension of the chains and running the simulations, the model was observed to have minimal angular rotation about the pivot point. At the same time, chain nodal oscillation and jerking was observed.

Chapter 3: Design Components

3.1 Selecting Chain Design

3.1.1 Options

A pool of possible designs was created to get a wide range of ideas. The designs were divided into 25 unique components focused on the structure of supports, types of caps, or the mobility of the system. The designs were individually evaluated and compared. Below are some of the examples of the designs compared.

The side block system

Shown in Figure 3.1.1.1. This is the system widely used in modern drydocking. This was considered the baseline for comparison.

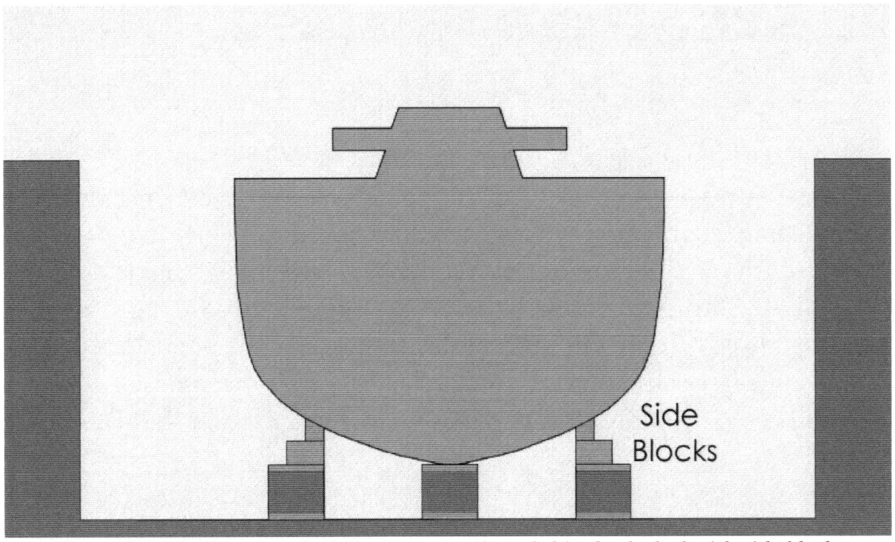

Figure 3.1.1.1: Diagram depicting cross section of ship drydocked with side blocks

The hauling block system

Shown in Figure 3.1.1.2, hauling blocks are another commonly used dry dock ship support method. In this system, the blocks have rounded caps (universal caps) and all are hauled in to contact the ship during the dry dock operation. This is accomplished as water is coming off the hull of this ship, after the ship has good contact with the keel blocks, but before the draft at instability. The system is underwater which causes problems with corrosion and visibility. These issues have resulted in the ship falling off the blocks. Another issue is the permanent rail installations that guide the hauling blocks limit the access to the ship.

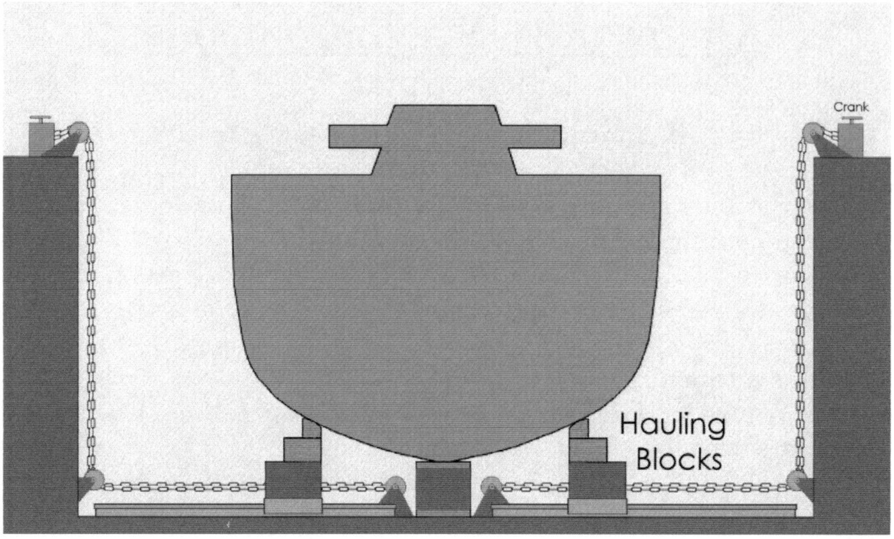

Figure 3.1.1.2: Diagram depicting cross section of ship drydocked with hauling blocks

Crank system

Shown in Figure 3.1.1.3, this design was explored for this project. The design features two design components, the "artillery cannon"-type mobility and crank structure design. The artillery cannon mobility system allows for the side blocks to be easily installed in location. This system was designed to be significantly lighter than the hauling block system, and may slide on the dry dock basin due to a decreased frictional force. The crank, similar to screw jack's system used on small boats, is cranked into the proper height. The cap is a universal cap just like the ones used on hauling blocks.

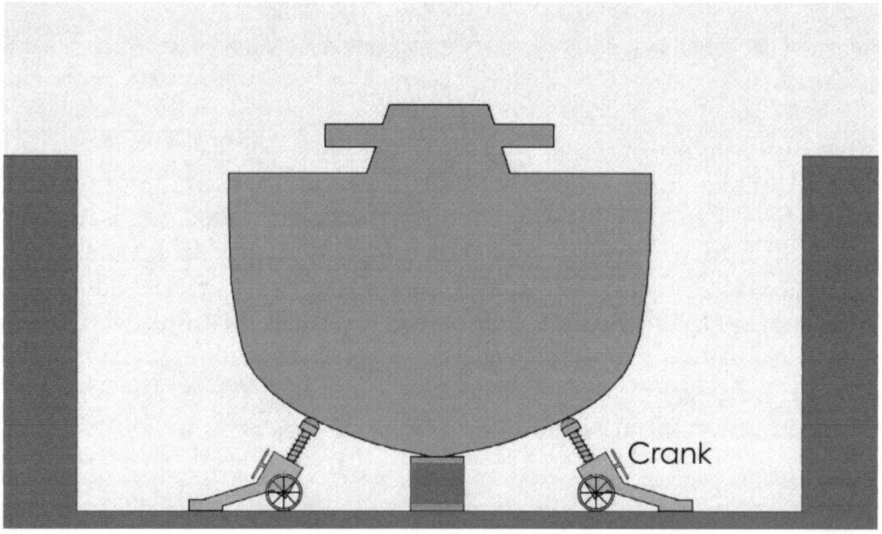

Figure 3.1.1.3: Diagram depicting cross section of a ship dry docked using the crank system

Man hoist system

Shown in Figure 3.1.1.4. The man hoist system uses the same artillery canon-type mobility system as seen in the crank system and uses the same universal caps. However, this system employs an alternate structure, based on man hoist machines. It was assumed that the since the man hoists already handle human life, the engineering behind them would be available and conservative.

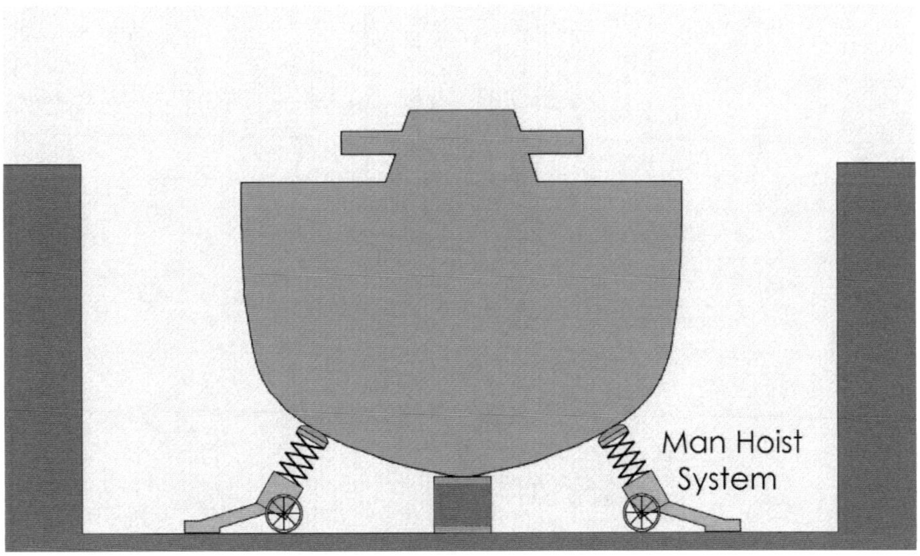

Figure 3.1.1.4: Diagram depicting cross section of a ship dry docked with the man hoist system

Binary system

Shown in Figure 3.1.1.6. The binary system uses the same artillery canon-type mobility system as seen in the crank or man hoist system and uses the same universal caps. This system uses a unique sectional design to create the structure of the support. The design is based off numbers in binary. Numbers in binary are made of a series of ones and zeros which signal signify the sum of a series of numbers in the base 2 exponential series (1 2 4 8 16...). In binary, the number 11 is the sum of 1, 2, and 8 (1 1 0 1). With this idea, sections of concrete of exponential length can be constructed (summed) to be set precisely to any specified length with only a few components.

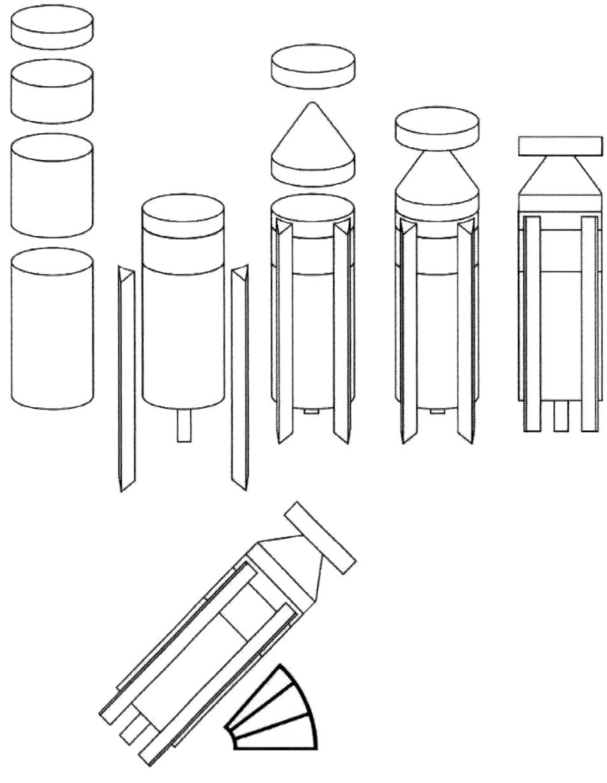

Figure 3.1.1.5: Diagram showing how exponentially larger sections can be used (just as in binary) to construct a support of any length

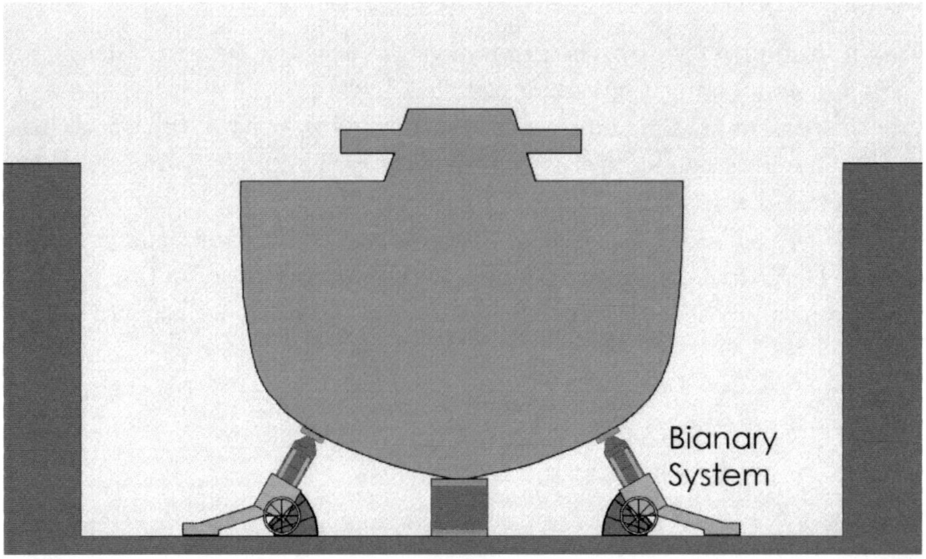

Figure 3.1.1.6: Diagram of a cross section of a ship dry docking using the binary system

Shores

Shown in Figure 3.1.1.7. Dry docking with shores is an antiquated process. Shown are whale shores which are installed as the ship is setting down on the keel block. This process needs to be done synchronously in the transverse direction, and is generally done as the ship is knuckling on the keel blocks. Shores have some history of falling into the basin, which can kill or injure personal and/or allow for the ship to fall off the blocks. Some dry dock yards tie the shores to the top of the dock to prevent them from falling into the basin. One advantage of shore is you have complete visibility of the system and of the contact with the hull.

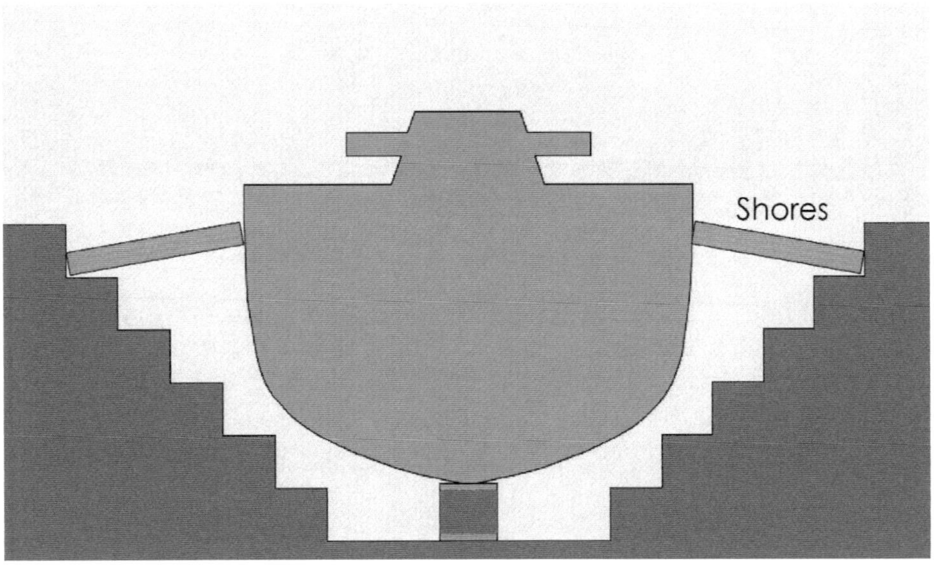

Figure 3.1.1.7: Diagram depicting a cross section of a ship dry docked with shores

Telescoping Shores

Shown in Figure 3.1.1.8. This design that was explored for this project, is as hydraulically powered version of shores. They have the same benefit as shores, being out of the water, where one can see the contact with the vessel. This system reduces the risk of shores by fixing them to the dock. The hydraulic would require some mechanical locking system. The expected upfront capital expense would be costly.

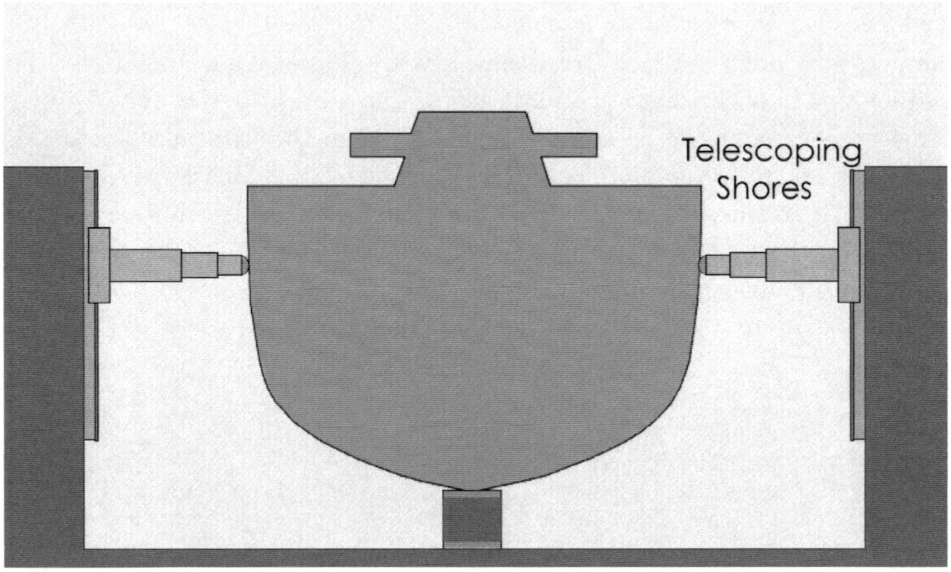

Figure 3.1.1.8: Diagram depicting a cross section of a ship dry docked with telescoping shores

Chains

Shown in Figure 3.1.1.9. Chains are not commonly used for drydocking. There is only one existing shipyard that was found to be using chains in conjunction with drydocking. Utilizing chains is a very efficient way of dry docking. The chain system is a very different system than any of the other systems used. Side blocks and shores support the ship on strength members in compression, while chains support the ship from bitts in tension.

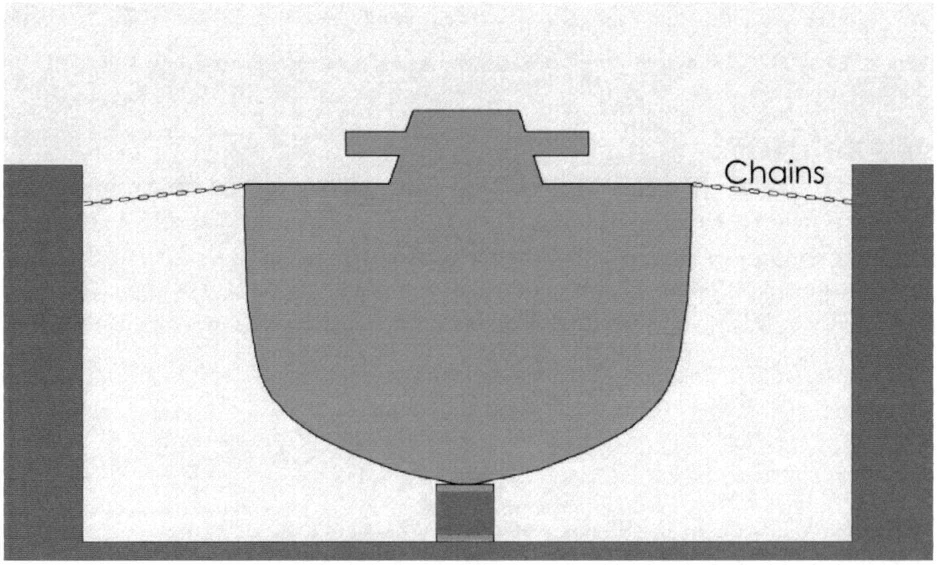

Figure 3.1.1.9: Diagram depicting a cross section of a ship dry docked with chains

Crane

Shown in Figure 3.1.1.10. The crane system was a design explored that is similar to the chain system. It added the ability for the chain connection point to be altered, although increased the initial cost of the system.

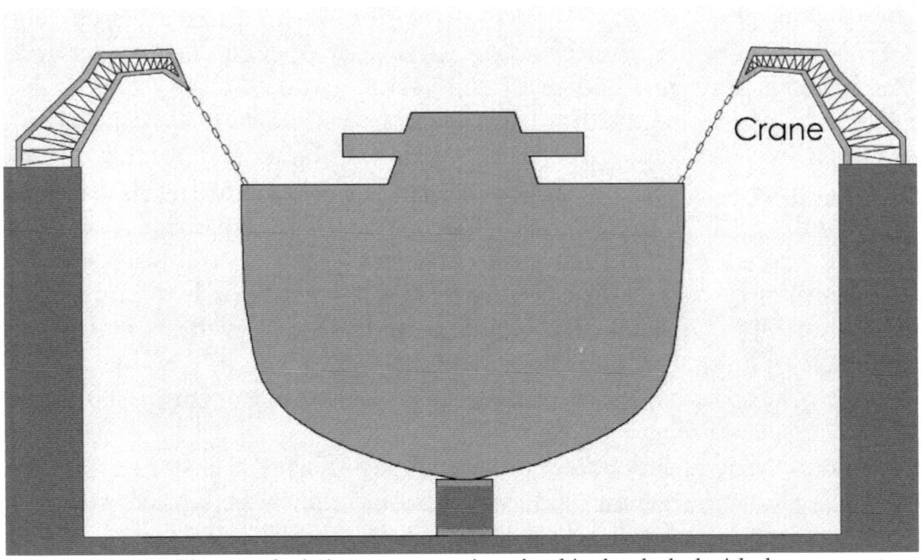

Figure 3.1.1.10: Diagram depicting a cross section of a ship dry docked with the crane system

3.1.2 Grading System

The designs were analyzed and the unique components categorized into separate design sections: The cap, the structure, and the mobility. Each component was given a name and discussed for advantages and disadvantages. Using the components of the possible designs along with input from the sponsor, a grading scale was created.

30% Cost	
100% overall	**20% Repair Accessibility**
initial	80% when active
maintenance	20% when stowed
lifetime	
man-hours	**10% Other**
set up	40% time to dry dock
operation	40% off centric loading
	15% visibility
30% Versatility	5% eco friendly
60% range	
33% long	**10% Integration**
33% trans	60% adaptability
33% vert	25% matching dp
20% hauling block type movement	15% simplistic
20% reinstall ability	

Figure 3.1.2.1: Design ranking system grading scale

Every possible design was graded based on several factors including, cost, versatility, repair accessibility, integration, and other minor considerations. Twenty-five possible designs were graded and ranked: See Figure 3.1.2.1.

The costs were estimated for each design based on estimates from the initial purchase and installation prices, the monetary cost of maintenance required, the life span of the system, and the man-hours required for set-up and during the operation.

The next category analyzed was the versatility of each design as a drydocking system. Each possible design had to fit the vessel as accurately as the side blocks. They needed mobility in longitudinal, transverse, and vertical directions. A system option for versatility was the ability for the system to be installed after the operation had already commenced. This procedure is like the hauling block system of drydocking. Hauling blocks are just like side blocks, although they are hauled out of the way while the ship is entering the dry dock, which allows for easier ship movement into the dry dock. The blocks are then hauled back into place against the ship's hull, allowing for a better fit. Finally, another option for versatility was the ability for blocks to be taken in and out while the ship was in dry dock. This ability allows for painting the underwater hull in spots that would normally be unreachable with the blocks in position against the hull.

Additionally, the repair accessibility was analyzed. If a system did not allow for ease of accessibility for repair, the system would not be feasible. This analysis for the system was directed towards supporting the ship while it was stowed. The 'other' category of the grading system included some miscellaneous items, the time spent in preparation to dry dock, could the system could support the ship if it were to be drydocked off centerline, visibility of components during certain stages of the operation, and its environmental impact.

The integration category was graded on the proposed system's adaptability to existing systems, comparison to the side blocks, and complexity.

Table 3.1.2.1: Design Rankings for All Unique Design Ideas

	Sidebloc	Shores	Hauling Bl	Hanging Si	Modified H	Yao Variat	Pin	Binary	Artillery Cr	Artillery Cr	Large Scre	Locker	Turn Lock
30% Cost	1.5	3	1.2	1.5	1.5	1.5	2.1	2.1	2.1	2.1	2.1	2.1	2.1
100% overall	5	10	4	5	5	5	7	7	7	7	7	7	7
30% Versatility	1.5	1.56	1.8	1.38	1.5	1.5	1.5	1.5	1.5	1.5	1.5	1.5	1.5
60% range	5	4	5	4.333333	5	5	5	5	5	5	5	5	5
33% long	5	4	5	5	5	5	5	5	5	5	5	5	5
33% trans	5	5	5	3	5	5	5	5	5	5	5	5	5
33% vert	5	3	5	5	5	5	5	5	5	5	5	5	5
20% hauling block type movement	5	5	10	5	5	5	5	5	5	5	5	5	5
20% reinstall ability	5	9	5	5	5	5	5	5	5	5	5	5	5
20% Repair Accessibility	1	2	0.92	1.96	0.56	1.16	1	1	1	1	1	1	1
80% when active	5	10	5	10	3	5	5	5	5	5	5	5	5
20% when stowed	5	10	3	9	2	9	5	5	5	5	5	5	5
10% Other	0.5	0.415	0.3	0.425	0.41	0.395	0.595	0.595	0.595	0.595	0.595	0.595	0.595
40% time to dry dock	5	5	5	7	7	7	7	7	7	7	7	7	7
40% off centric loading	5	1	0	0	0	0	5	5	5	5	5	5	5
15% visibility	5	10	5	7	6	5	5	5	5	5	5	5	5
5% eco friendly	5	5	5	8	8	8	8	8	8	8	8	8	8
10% Integration	0.5	0.355	0.38	0.455	0.335	0.44	0.5	0.5	0.5	0.5	0.5	0.5	0.5
60% adaptability	5	3	3	5	3	4	5	5	5	5	5	5	5
25% matching dp	5	1	5	5	5	5	5	5	5	5	5	5	5
15% simplistic	5	10	5	2	2	5	5	5	5	5	5	5	5
TOTAL	50	73.3	46	57.2	43.05	49.95	56.95	56.95	56.95	56.95	56.95	56.95	56.95

	Tripod	Man Lift	Manlift Sin	Simple Hydr	Double Hydra	Supported	Chain	Bridge	Crane	Telescopin	Hydraulic	Sideblocks
30% Cost	1.5	1.5	1.5	1.5	1.5	1.5	2.4	1.5	1.5	1.5	1.5	1.5
100% overall	5	5	5	5	5	5	8	5	5	5	5	5
30% Versatility	1.44	1.86	1.86	1.86	1.86	1.86	2.1	1.86	1.98	1.74	1.74	1.5
60% range	4.666667	5	5	5	5	5	5	5	5	4.333333	4.333333	5
33% long	5	5	5	5	5	5	5	5	5	4	4	5
33% trans	4	5	5	5	5	5	5	5	5	4	4	5
33% vert	5	5	5	5	5	5	5	5	5	5	5	5
20% hauling block type movement	5	8	8	8	8	8	10	8	9	8	8	5
20% reinstall ability	5	8	8	8	8	8	10	8	9	8	8	5
20% Repair Accessibility	0.68	1	1	1	1	1	2	2	2	1.96	1.96	1
80% when active	3	5	5	5	5	5	10	10	10	10	10	5
20% when stowed	5	5	5	5	5	5	10	10	10	9	9	5
10% Other	0.435	0.595	0.595	0.595	0.595	0.595	0.67	0.67	0.63	0.55	0.55	0.5
40% time to dry dock	7	7	7	7	7	7	9	9	9	9	9	5
40% off centric loading	1	5	5	5	5	5	3	3	2	0	0	5
15% visibility	5	5	5	5	5	5	10	10	10	10	10	5
5% eco friendly	8	8	8	8	8	8	8	8	8	8	8	5
10% Integration	0.485	0.485	0.485	0.485	0.485	0.485	0.33	0.27	0.285	0.18	0.18	0.5
60% adaptability	5	5	5	5	5	5	4	4	4	2	2	5
25% matching dp	5	5	5	5	5	5	0	0	0	0	0	5
15% simplistic	4	4	4	4	4	4	6	2	3	4	4	5
TOTAL	45.4	54.4	54.4	54.4	54.4	54.4	75	63	63.95	59.3	59.3	50

Each category and subcategory had a possible score of 0-10. To baseline the grading scale, the existing system of side blocks was given a base score of 5 in all categories. Each possible design was given a score relative to the existing design. From the results of the ranking system, some notable unique designs were chosen and reanalyzed for advantages and disadvantages. Eventually, chains were the evaluated as the top scoring design choice. See Figure 3.1.3.1.

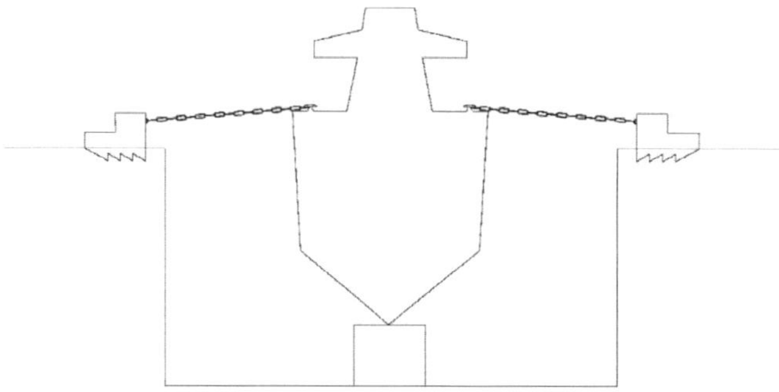

Figure 3.1.2.2: Initial chain concept diagram

3.1.3 Design Choice

The system of chains was chosen as the most optimal design. Chains did well in all categories of the grading system except for the integration category. See Table 3.1.2.1. This system is very different from the existing systems of drydocking, which poses a possible issue in marketing the system. We also noted that this system had a single component failure. If any one component of the chain system, including a single chain link failed, this would cause a possible failure of the system. Adding sufficient number chains to the system lowered the failure rate to an acceptable level. The system has an optimistic cost analysis, good versatility, the best repair accessibility, complete visibility and conveniently quick operational time.

3.2 Static Analysis

3.2.1 Purpose

The static analysis was based on an applied force on the ship. The concern is the ship overturning or slipping off the blocks in the transverse direction (port and starboard). See Figure 3.2.1.1.

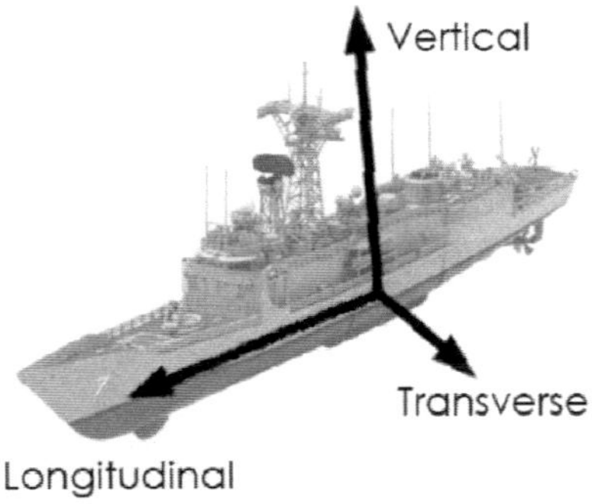

Figure 3.2.1.1: Various directions for ships

The analysis used the forces derived for earthquakes and winds from standard US Navy calculations to determine the resultant forces of tension in the chains, force of the ship on the keel blocks and the friction coefficient required to keep the ship from slipping off the keel blocks.

3.2.2 Applied Forces Derivation

There are two excitation forces that the US Navy preforms (Ref 1) for static loading of the ship while it sits on the blocks. The two cases are force from an earthquake and force from a high wind.

The calculations as seen in the industry are a pressure balance as seen in equation 3.2.2.1.

$$\frac{\frac{M}{l_2}+\Delta\times\frac{15\%}{2}}{A_2\times N_b} = P_f \qquad \text{(Eq 3.2.2.1)}$$

M= Moment of the applied force
l_2= Average half breadth (transverse) length of each side block from the keel
Δ= Displacement of ship (weight of ship)
A_2= Contact area of each side block to the hull
N_b= Number of side blocks per side
P_f= Yield pressure of softwood cap

Side blocks are composed of concrete, hardwood, and softwood. Concrete and hardwood have a much higher yield strength, so the softwood determines the weak link in the system and therefore the focus of analysis. The calculations test to ensure the block's softwood caps are not overloaded. See Figure 3.2.2.1.

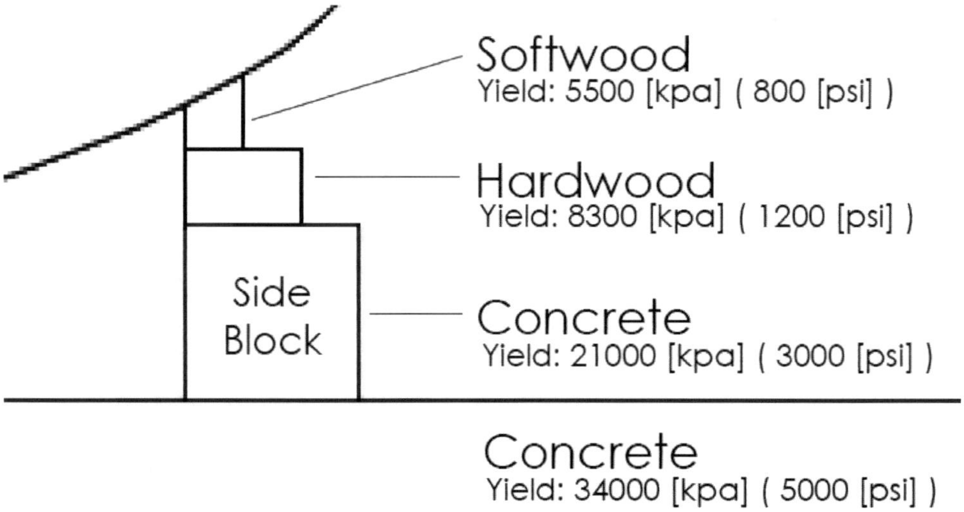

Figure 3.2.2.1: Graphics showing lower failure yield limit of side block's softwood cap

Equation 3.2.2.1 can be rearranged to represent a moment balance as seen in equation 3.2.2.2.

$$M + \Delta \times \frac{15\%}{2} \times l_2 = P_f \times A_2 \times N_b \times l_2 \qquad \text{(Eq 3.2.2.2)}$$

In equation 3.2.2.2 the left side of the equation is the moment from the earthquake and hurricane added to the "dead load". The dead load is the approximate amount of the ship's weight that the side blocks are supporting with no earthquake or wind forces applied. This value is approximated to 15% of the ship weight total, divided by two to represent the weight carried by one side.

Equation 3.2.2.3 represents the moment for the earthquake equation. This was based off a 0.2 g earthquake. This is a significant earthquake and classified as a category XII earthquake on the Richter Scale.

$$M = \Delta \times KG \times 0.2 \qquad \text{(Eq 3.2.2.3)}$$

KG= Distance Keel to Center of gravity

Equation 3.2.2.4 represents the moment for the wind equation.

$$M = 0.19 \times V^2 \times \Sigma(A \times l) \qquad \text{(Eq 3.2.2.4)}$$

The moments can then be transformed from applied movements on the bodies to applied accelerations simply as seen in equation 3.2.2.5

$$a = \frac{M}{m \times KG} \qquad \text{(Eq 3.2.2.5)}$$

a= Applied acceleration
m= Mass of the ship

3.2.3 Diagrams

The geometry of the ship is labeled in Figure 3.2.3.1.

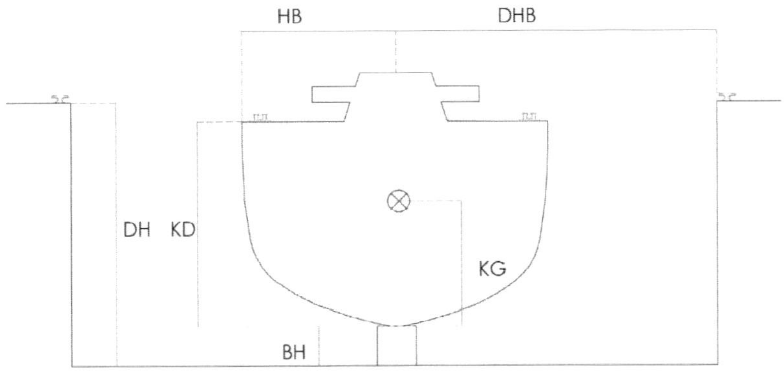

Figure 3.2.3.1: Diagram labeling important geometric parameters of ship and dry dock

KG= Distance Keel to Center of gravity
KD= Distance Keel to Main deck
HB= Half breadth of ship
BH= Block Height
DH= Dock Height
DHB= Dock Half-Breadth

The free body diagram of the ship was developed for static analysis. See Figure 3.2.3.2.

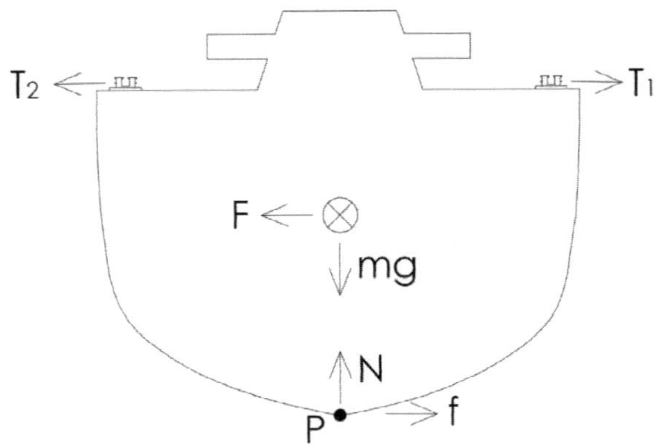

Figure 3.2.3.2: Free body diagram of the vessel

T_1= Tension from chain 1 (right side)
T_2= Tension from chain 2 (left side)
F= Force applied to the body
f= Force of friction between ship and the keel blocks
N= Normal force
m= Mass of the ship
g= Gravity constant

P= Pivot point

Note the angle of Tension in Figure 3.2.3.2 is just shown as horizontal. The tension will not be horizontal in practice. The angle will be taken at several values throughout the analysis and divided into components: the x and y components and the component that applies to the overturning moment. The component that applies to the overturning moment is calculated through angles shown in Figure 3.2.3.3.

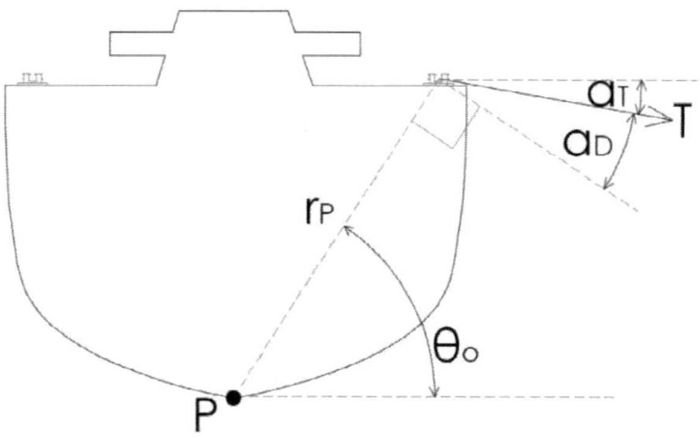

Figure 3.2.3.3: Diagram labeling important angle and geometric parameters relating to a chains tension angle and the righting moment of the ship

α_T= Absolute angle of chain
θ_o= Angle of radius from point P to chain attachment
α_D= Divergent angle of chain from contributing to counter moment from point P
r_P = Lever arm from chain attachment to point P

From these diagrams the static equations were developed.

3.2.4 Equations

Geometry

Divergent angle of chain from contributing to counter moment from point P

$$\alpha_D = 90° - \theta_o - \alpha_T \qquad \text{(Eq 3.2.4.1)}$$

Lever arm from chain attachment to point P

$$r_P = \sqrt{KD^2 + HB^2} \qquad \text{(Eq 3.2.4.2)}$$

Force vector calculations

Vector force of chain 1 tension in x direction

$$T_{P1x} = T_1 \times \cos\theta_o \qquad \text{(Eq 3.2.4.3)}$$

Vector force of chain 1 tension in y direction

$$T_{P1y} = T_1 \times \sin\theta_o \qquad \text{(Eq 3.2.4.4)}$$

Vector force of chain 1 tension contributing to counter moment from point P

$$T_{P1} = T_1 \times \cos\alpha_D \qquad \text{(Eq 3.SA.5)}$$

Vector force of chain 2 tension in x direction

$$T_{P2x} = T_2 \times \cos\theta_o \qquad \text{(Eq 3.2.4.6)}$$

Vector force of chain 2 tension in y direction

$$T_{P2y} = T_2 \times \sin\theta_o \qquad \text{(Eq 3.2.4.7)}$$

Vector force of chain 2 tension contributing to counter moment from point P

$$T_{P2} = T_2 \times \cos\alpha_D \qquad \text{(Eq 3.2.4.8)}$$

Force equations

Force of friction between ship and the keel blocks

$$f = \mu N \qquad \text{(Eq 3.2.4.9)}$$

where: μ= friction coefficient between the keel and the keel blocks

Force applied to body

$$F = ma \qquad \text{(Eq 3.2.4.10)}$$

where: a= acceleration applied to body

System of equations

Sum of the forces in the X direction taking left to be positive.

$$\leftarrow+ \quad \Sigma F_x: F - f - T_{P1x} + T_{P2x} = 0 \qquad \text{(Eq 3.2.4.11)}$$

Sum of the forces in the Y direction taking up to be positive.

$$\uparrow+ \quad \Sigma F_y: N + T_{P1y} + T_{P2y} - mg = 0 \qquad \text{(Eq 3.2.4.12)}$$

Sum of the moments about point P taking counter-clockwise to be positive.

$$\circlearrowleft+ \quad \Sigma M_P: KG \times F - r_P \times T_{P1} + r_P \times T_{P2} = 0 \qquad \text{(Eq 3.2.4.13)}$$

3.2.5 Parameters

KG=5.79 [m] (19 [ft]) Distance Keel to Center of gravity (Must be provided by ship's loading books, this value provided by typical value for FFG-7 given in Ref 1)

KD=9.44 [m] (31 [ft]) Distance Keel to Main deck (Ref 2)

HB=7.06 [m] (23.5 [ft]) Half breadth of ship (Ref 2)

BH=1.83 [m] (6 [ft]) Block Height (Ref 3)

DH=12.19 [m] (40 [ft]) Dock Height (Ref 3)

DHB=15.24 [m] (50 [ft]) Dock Half-Breadth (Ref 3)

3.2.6 Results

For this system of equations there are 4 variables (T1, T2, μ, N) and only 3 equations (3.2.4.11-13). In order to complete calculations for this static position one variable needed to be eliminated. T2 was assumed to be zero for this analysis. The results from solving this system of equations were:

T1 = 4.71×10^6 [N] = 472 [LT]

This was an approximation for the righting tension required for the applied load on the ship. A more detailed analysis will be covered in sections 3.3.

N = 3.45×10^7 [N] = 3450 [LT]

This was a good estimate for the normal force applied by the blocks on the ship. This force could then be used to check the loading on the blocks. It is important to note that by eliminating the side blocks, the keel blocks would feel a greater force. It was estimated that the side blocks contribute to 15% of the ship's displacement out of water (Ref 1). If loading exceeds the safe working load for the keel blocks, more keel blocks should be added to compensate.

$\mu = 0.0672$

This result suggests the ship will not slip off the blocks. The coefficient of friction between wood and steel was expected to be 0.15-0.25 depending on the level of the wood moisture (Ref 8).

3.3 Dynamic Analysis

3.3.1 Purpose

The static analysis sets up a good estimate for the forces required for maximum stability. The dynamic analysis should more accurately describe the system by introducing an angle of rotation to the system. For this analysis, it will be assumed that the vessel is pinned at the keel and rotating transversely. Additionally, the dynamics of the chain will also be considered. The chain will change tension based on the same rotation of the ship. The section will again analyze the slip conditions, block loading and tension in the chains.

3.2.2 Diagrams

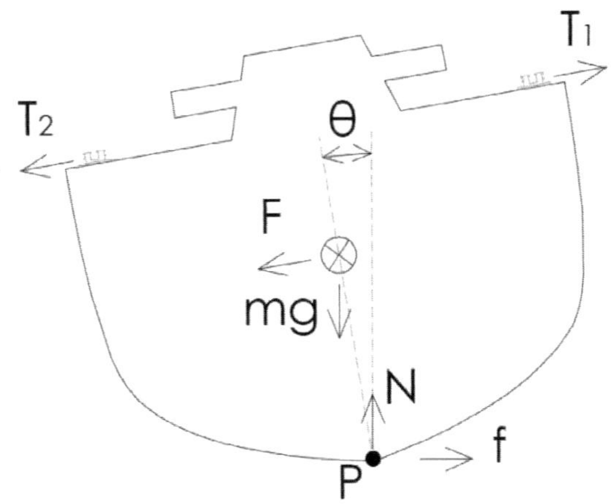

Figure 3.2.2.1: Free body diagram of the vessel rotating about point P (reference Figure 3.2.3.2)

θ= Angle of rotation

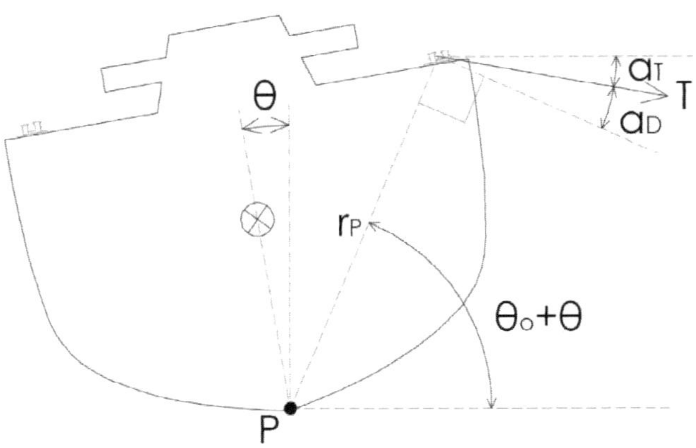

Figure 3.2.3.3: Diagram labeling important angle and geometric parameters relating to a chains tension angle and the righting moment of the ship as the vessel is rotating about point P (reference Figure 3.3.3.3)

3.3.3 Equations

The weight affects chain tension versus angle. Due to these changes from the static analysis to the dynamic analysis, some of the static equations must be updated from 3.2.1.

Geometry

Divergent angle of chain from contributing to counter moment from point P.

$$\alpha_D = 90° - (\theta_o + \theta) - \alpha_T \qquad \text{(Eq 3.3.3.1)}$$

Force vector calculations

Vector force of chain 1 tension in x direction

$$T_{P1x} = T_1 \times \cos(\theta_o + \theta) \qquad \text{(Eq 3.3.3.2)}$$

Vector force of chain 1 tension in y direction

$$T_{P1y} = T_1 \times \sin(\theta_o + \theta) \qquad \text{(Eq 3.3.3.3)}$$

Vector force of chain 2 tension in x direction

$$T_{P2x} = T_2 \times \cos(\theta_o + \theta) \qquad \text{(Eq 3.3.3.4)}$$

Vector force of chain 2 tension in y direction

$$T_{P2y} = T_2 \times \sin(\theta_o + \theta) \qquad \text{(Eq 3.3.3.5)}$$

Force vector calculations for applied force in x direction

$$F_x = F \times \cos\theta \qquad \text{(Eq 3.3.3.6)}$$

Force vector calculations for applied force in y direction

$$F_y = -F \times \sin\theta \qquad \text{(Eq 3.3.3.7)}$$

System of equations

Sum of the forces in the X direction taking left to be positive.

$$\leftarrow + \qquad \Sigma F_x: F_x - f - T_{P1x} + T_{P2x} = 0 \qquad \text{(Eq 3.3.3.8)}$$

Sum of the forces in the Y direction taking up to be positive.

$$\uparrow + \qquad \Sigma F_y: -F_y + N + T_{P1y} + T_{P2y} - mg = 0 \qquad \text{(Eq 3.3.3.9)}$$

Sum of the moments about point P taking counter-clockwise to be positive.

$$\curvearrowleft + \qquad \Sigma M_P: KG \times F - r_P \times T_{P1} + r_P \times T_{P2} + mg \times KG \times \sin(\theta) = 0 \qquad \text{(Eq 3.3.3.10)}$$

3.3.4 Catenary Analysis

The chain will sag when stretched horizontally, even if only a small amount. The curve created by this sag makes a catenary. The catenary equations can describe the shape of a chain, while also providing the tension in the chain based on the chain's force per unit length. To model it, the chain must be assumed to have infinitesimally small links and no elasticity.

3.3.5 Catenary Diagram and Equations

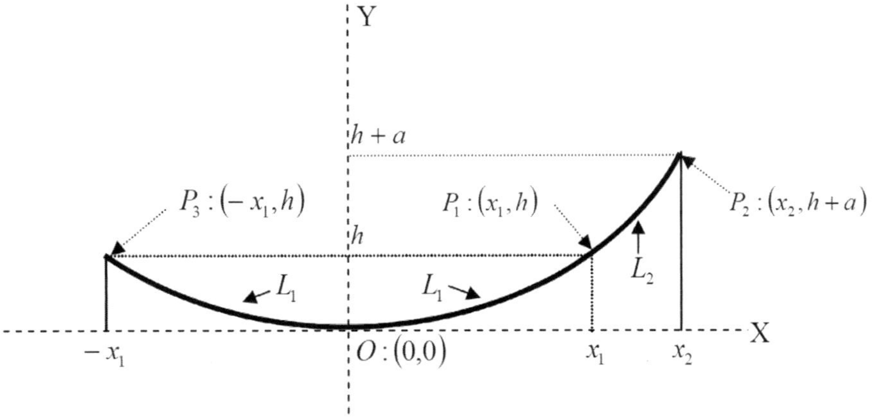

Figure 3.3.5.1: Diagram labeling many important parameters of a catenary curve (Ref 2)

h= Sag of the chain
a= Difference in heights of the chain attachment points
P_1= Arbitrary point between P_2 and P_3 where height is equal
P_2= Chain attachment to ship
P_3= Chain attachment to dock
x_1= Horizontal distance from lowest point to chain attachment to dock
x_2= Horizontal distance from P_2 to P_3
L_1= Length of chain from lowest point to attachment on dock
L_2= Length from P_1 to P_2

The first equation represents whether the chain is geometrically over constrained. If the rope's length is shorter than the distance between the two attachment points, or if l does not satisfy Eq 3.3.5.1, then the assumptions have been violated and the elasticity of the chain must be considered. This is not covered in the chain specifications.

$$l \geq \sqrt{a^2 + d^2} \qquad \text{(Eq 3.3.5.1)}$$

where l = total length of the chain
d = Horizontal distance between the two attachment points.

Another parameter of the geometry of the chain must be checked. The case in which there is no sag between the two points on the catenary.

$$\frac{l^2 - a^2}{2a} \ln\left(\frac{l+a}{l-a}\right) > d \qquad \text{(Eq 3.3.5.2)}$$

If Eq 3.3.5.2 is satisfied then Eq 3.3.5.3-4 are used.

$$d = \frac{(l^2-a^2)(a+2h) - 2 \times l\sqrt{h(a+h)(l^2-a^2)}}{a^2} \operatorname{atanh}\left(\frac{a^2}{l(a+2h) - 2\sqrt{h(a+h)(l^2-a^2)}}\right) \qquad \text{(Eq 3.3.5.3)}$$

$$l_1 = -\frac{h \times l - \sqrt{h(a+h)(l^2-a^2)}}{a} \qquad \text{(Eq 3.3.5.4)}$$

If Eq 3.3.5.2 is not satisfied then Eq 3.3.5.5-6 are used.

$$d = \frac{(l^2-a^2)(a+2h)+2\times l\sqrt{h(a+h)(l^2-a^2)}}{a^2}\text{atanh}\left(\frac{a^2}{l(a+2h)+2\sqrt{h(a+h)(l^2-a^2)}}\right) \quad \text{(Eq 3.3.5.5)}$$

$$l_1 = -\frac{h\times l+\sqrt{h(a+h)(l^2-a^2)}}{a} \quad \text{(Eq 3.3.5.6)}$$

Eq 3.3.5.3 or Eq 3.3.5.5 is used to numerically solve for h. There is no direct formula to solve for h.

$$\mu = \frac{lh}{(l_1^2-h^2)} \quad \text{(Eq 3.3.5.7)}$$

where μ= A dimensionless parameter of the catenary

Catenary equation

$$y = \frac{\cosh(\mu\times x)}{\mu} \quad \text{(Eq 3.3.5.9)}$$

where x and y are Cartesian variables that describe the curve of the catenary Additional parameters can be calculated to that can be used to set the

$$x_1 = \frac{\text{asinh}(\mu\times l_1)}{\mu} \quad \text{(Eq 3.3.5.10)}$$

$$k = \frac{\cosh(\mu(-(d-x_1)))}{\mu} \quad \text{(Eq 3.3.5.11)}$$

where k= Catenary vertical adjustment parameter

Adjusted catenary equation

$$y = \frac{\cosh(\mu(x-(d-x_1)))}{\mu} - k \quad \text{(Eq 3.3.5.12)}$$

Tension calculation

$$Td = \lambda\times g\times\frac{{l_1}^2+h^2}{2h} \quad \text{(Eq 3.3.5.13)}$$

where *Td*=Tension in chain at the dock side chain attachment

$$Ts = \lambda\times g\times\frac{{l_1}^2+h^2}{2h} + \frac{\lambda\times g\left(\frac{(l+l_1)^2+(h+a)^2}{2(h+a)}-\frac{l^2+h^2}{2h}\right)}{2} \quad \text{(Eq 3.3.5.14)}$$

where *Ts*= Tension in the chain at the ship side chain attachment

$$\omega_{Td} = \sinh(\mu\times x_1) \quad \text{(Eq 3.3.5.15)}$$

where ω_{Td}= Angle of the chain from the horizontal of the attachment to the dock

$$\omega_{Ts} = \sinh(\mu \times (d - x_1)) \qquad \text{(Eq 3.3.5.16)}$$

where ω_{Td}= Angle of the chain from the horizontal of the attachment to the ship

A comparison can be made evaluate which side of the dock is higher. In the case where the left side is higher, Eq 3.3.5.13 would solve for the tension at on the ship and Eq 3.3.5.14 would solve for the tension on the dock. Eq 3.3.5.15 would solve for the angle of the chain from the horizontal of the attachment to the ship and Eq 3.3.5.16 would solve for the angle of the chain from the horizontal of the attachment to the dock.

For equation derivations see Ref 2.

Sum of the forces in the X direction taking left to be positive.

$$\leftarrow + \quad \Sigma F_x: F_x - f - T_{P1x} + T_{P2x} = 0 \qquad \text{(Eq 3.3.3.8)}$$

Sum of the forces in the Y direction taking up to be positive.

$$\uparrow + \quad \Sigma F_y: -F_y + N + T_{P1y} + T_{P2y} - mg = 0 \qquad \text{(Eq 3.3.3.9)}$$

Sum of the moments about point P taking counter-clockwise to be positive.

$$\circlearrowleft + \quad \Sigma M_P: KG \times F - r_P \times T_{P1} + r_P \times T_{P2} + mg \times KG \times \sin(\theta) = 0 \qquad \text{(Eq 3.3.3.10)}$$

But now T1 and T2 are a function of theta

Longer length with same distance vs shorter length with same distance

Increases tension

Longer distance vs shorter distance with same tension at same theta find lower Tension at 0

Lower at shorter lengths. Under the same tension, it would sag more so the non-linear trend would be less profound. Could optimize the position to have the shortest chains possible.

3.3.6 Catenary Optimization

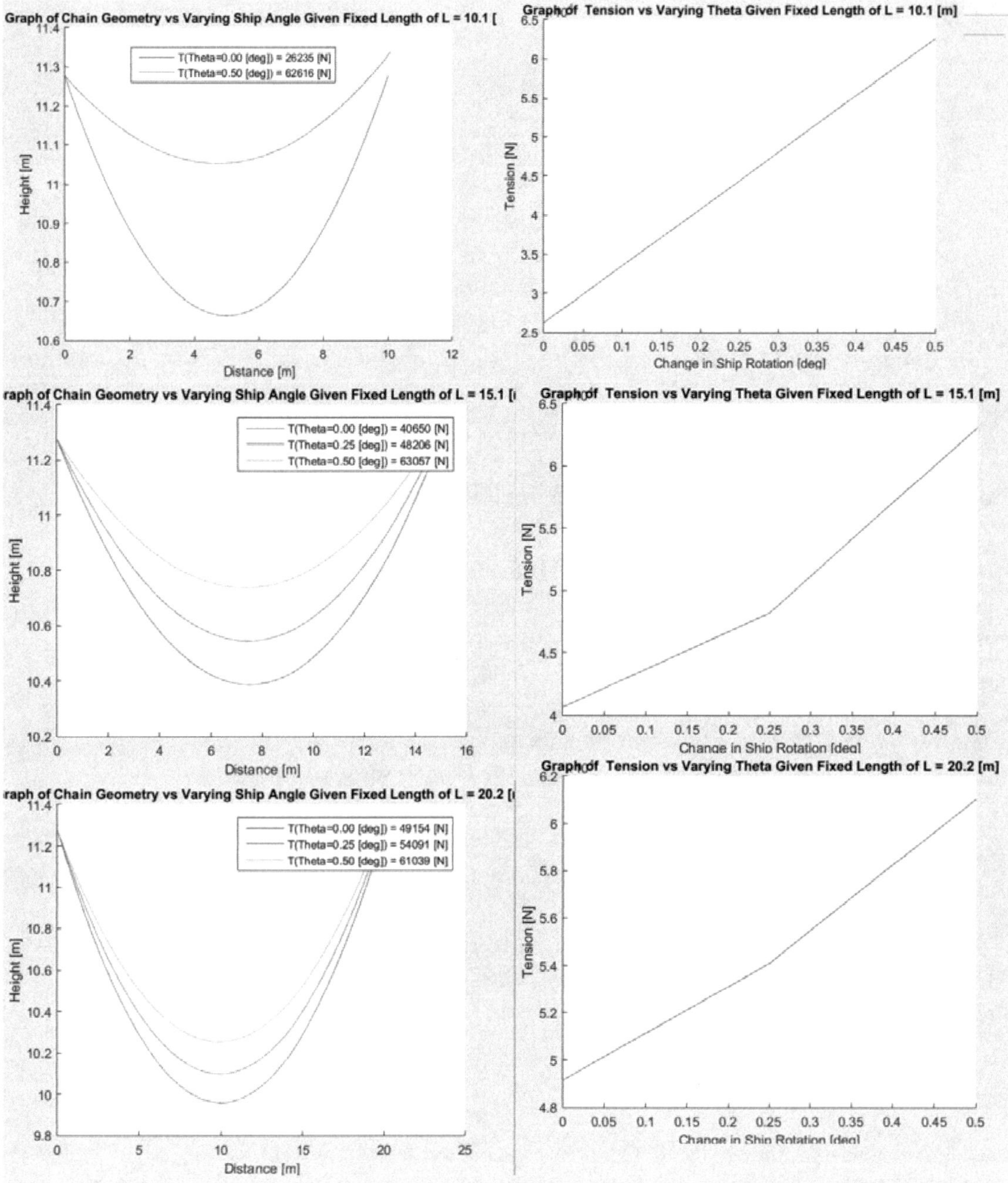

Figure 3.3.6.1: Graphs of catenary results for showing steeper graphs for shorter distances

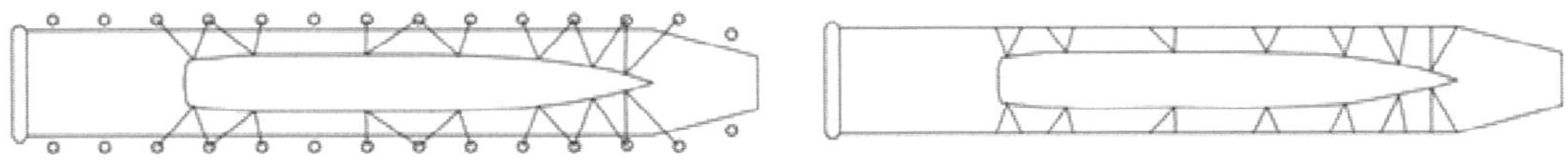

Figure 3.3.6.2: Diagrams of longitudinal attachment and system placement

Angled vs horizontal:

There is almost no effect for the tested angles, but flatter is better. A vertical chain will have no catenary. This angle could be optimized against the divergent angle for the perfect chain fit.

Slip dynamic trial 1

Table 3.3.6.1: Theta vs Tension

Theta [deg]	Tension [N]	Angle of dock side attachment [deg]	Angle of ship side attachment [deg]
-0.5	15080.94	17.12115	22.82145
-0.4	15835.03	16.03182	21.93034
-0.3	16730.06	14.87897	20.97769
-0.2	17815.01	13.65078	19.9517
-0.1	19166.44	12.33105	18.8362
0	20912.32	10.89651	17.60794
0.1	23286.97	9.311526	16.23132
0.2	26780.45	7.516254	14.64651
0.3	32663.38	5.394293	12.73716
0.4	45988.4	2.645978	10.20361
0.5	1141271	-3.62898	4.145613

Plugging into dynamic system of equations:

Inputs:

Theta=0 [deg]

F1= 20912 [N]

F2=20912 [N]

Outputs:

N= 3.4969e+07 [N]

μ = 0.2000

Inputs:

Theta=0.1 [deg]

F1=23286.97

F2=19166.44

Outputs:

N= 3.4969e+07 [N]

μ = 0.1999

Inputs:

Theta=0.2 [deg]

F1= 26780

F2= 17815

Outputs:

N= 3.4969e+07 [N]

μ = 0.1998

Inputs:

Theta=0.3 [deg]

F1=32663.38

F2=16730.06

Outputs:

N= 3.4968e+07 [N]

μ = 0.1996

Inputs:

Theta = 0.5 [deg]

F1=425380.7

F2=15080.94

Outputs:

N= 3.4925e+07 [N]

μ = 0.1886

The system requires the highest coefficient of friction which is at theta=0. This is because forces cancel in horizontal direction allowing for slip in a condition where the earthquake is above a 0.2gs. This can be mitigated with use of rubber caps which would increase the coefficient of friction. Alternatively spur shore type stabilizers can added once the ship is in dock.

As proof of concept:

Inputs:

Theta=0 [deg]

F1= 20912

F2=20912

For 1g earthquake

Outputs:

N= 3.4969e+07 [N]

μ = 0.1000

Table 3.3.6.1: Full Table of Theta vs Tension

Theta [deg]	Tension [N]	Angle of dock side attachment [deg]	Angle of ship side attachment [deg]
-0.5	15080.94	17.12115	22.82145
-0.4	15835.03	16.03182	21.93034
-0.3	16730.06	14.87897	20.97769
-0.2	17815.01	13.65078	19.9517
-0.1	19166.44	12.33105	18.8362
0	20912.32	10.89651	17.60794
0.1	23286.97	9.311526	16.23132
0.2	26780.45	7.516254	14.64651
0.3	32663.38	5.394293	12.73716
0.4	45988.4	2.645978	10.20361
0.4	45988.4	2.645978	10.20361
0.41	48452.08	2.307145	9.886373
0.42	51364.64	1.949354	9.550201
0.43	54880.98	1.568942	9.191429
0.44	59242.86	1.160878	8.805028
0.45	64853.6	0.717931	8.383766
0.46	72450	0.229016	7.916557
0.47	83568.17	-0.32449	7.384781
0.48	102171.6	-0.97878	6.752244
0.49	143859.3	-1.82664	5.926153
0.49	143859.3	-1.82664	5.926153
0.491	151501.3	-1.93117	5.823806
0.492	160508.3	-2.04152	5.715635
0.493	171341.9	-2.1588	5.600533
0.494	184716.5	-2.28452	5.476986
0.495	201810.8	-2.42085	5.342835
0.496	224745.8	-2.57107	5.194804
0.497	257842.9	-2.74057	5.027475
0.498	311825.8	-2.93964	4.830591
0.499	425380.7	-3.19314	4.579272
0.5	1141271	-3.62898	4.145613

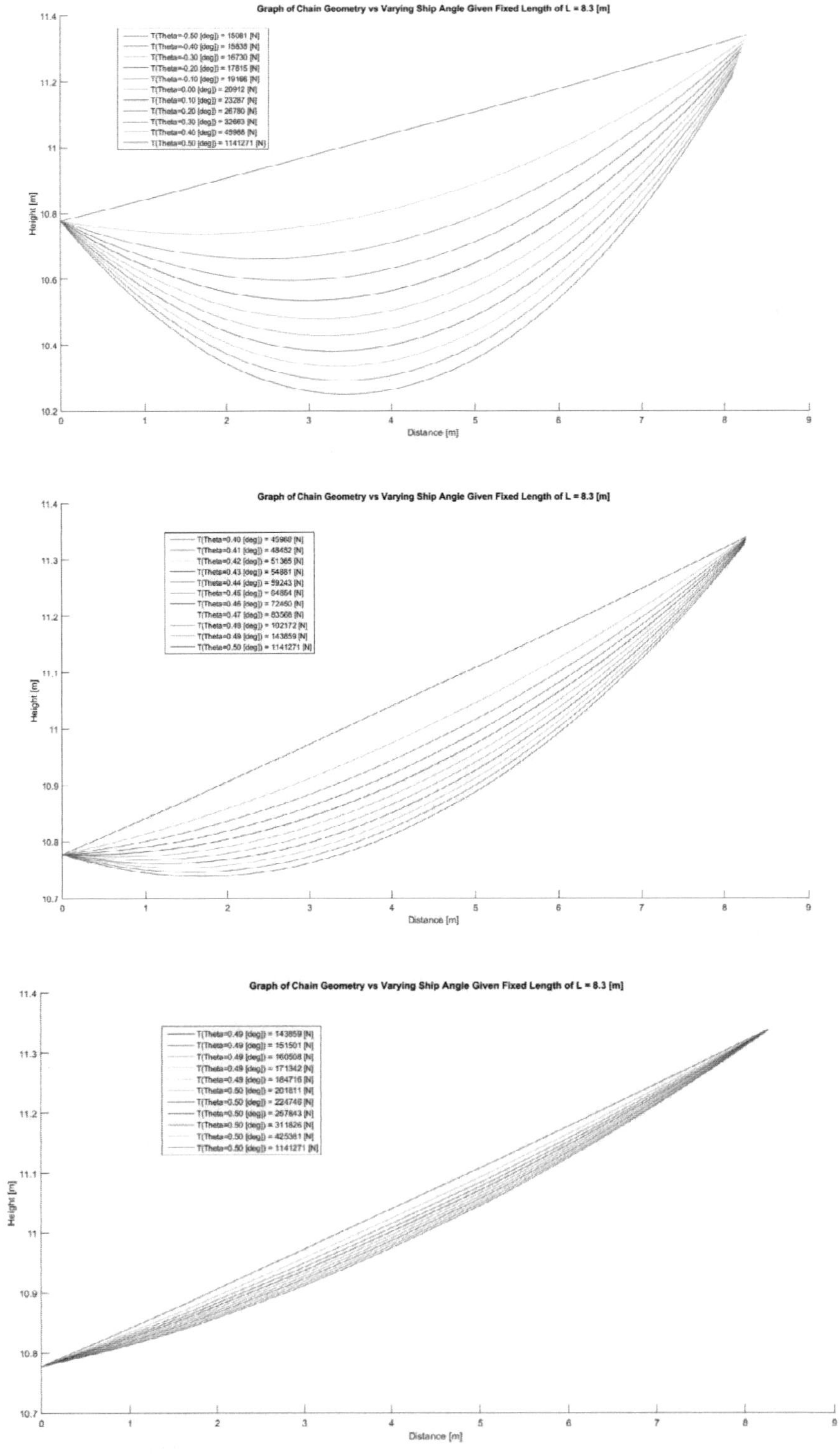

Figure 3.3.6.3: Graphs of catenary results showing extreme jerking force as chain becomes taught

Optimization resulting in the least linear curve is best with the shortest distance between chain attachments, and the least height differential between chain attachments (although this is not as important). Chain weight was not factored in the calculations.

3.4 Vibrations Analysis

3.4.1 Analysis

We need to prove that the non-linear behavior of the catenary will eliminate the resonance frequency such that an earthquake will not excite the system. The standard formulas for vibrations analysis are as following:

$$I\ddot{\theta} + c_R\dot{\theta} + k_R\theta = \tau(\theta) \qquad \text{(Eq 3.4.1.1)}$$

$$m\ddot{x} + c\dot{x} + kx = F(x) \qquad \text{(Eq 3.4.1.2)}$$

$$\theta = \frac{x}{r}, \dot{\theta} = \frac{\dot{x}}{r}, \ddot{\theta} = \frac{\ddot{x}}{r} \qquad \text{(Eq 3.4.1.3)}$$

$$k_R = r^2 k \qquad \text{(Eq 3.4.1.4)}$$

$$c_R = r^2 c \qquad \text{(Eq 3.4.1.5)}$$

$$I = I_{cg} + mr^2 \qquad \text{(Eq 3.4.1.6)}$$

I_{cg} is inertia of the ship

Our system is different. It includes a 2nd order non-linear ODE that also requires numerical integration. We have catenary equations inside the ODE that give us issues. We also will look at damping from friction in chain links and elasticity of the chain. The effect on gravity due to the system rotation was also considered.

Damping between chain links:

Energy dissipation between chain links

$$E_c = \gamma(\theta) \times r_c \times \mu_2 \times T(\theta) \qquad \text{(Eq 3.4.1.7)}$$

where γ=Total change in angle of the links, r_c= radius of the chain link, μ_2= coefficient of friction between two chain links, T= tension in the chain

Figure 3.4.1.1: Diagram illustrating the change in angles resulting in gamma

Damping dissipation formula

$$E_D = D\times\dot{\theta}\times\Delta s \qquad \text{(Eq 3.4.1.8)}$$

where D= damping coefficient and Δs= change in distance

$$\Delta s = \sqrt{\left(DBH - HB + r_p\times\cos(\theta)\right)^2 + \left(DH - (BH + KD) + r_p\times\sin(\theta)\right)^2} - \sqrt{(DBH - HB)^2 + \left(DH - (BH + KD)\right)^2} \qquad \text{(Eq 3.4.1.9)}$$

Energy balance of the two $E_C = E_D$ and solving for c

$$D(\theta) = \frac{\gamma(\theta)\times r_C\times\mu_2\times T(\theta)}{\dot{\theta}\times\Delta s} \qquad \text{(Eq 3.4.1.10)}$$

Elasticity of the chain:

$$E_1(\theta) = E\times\Delta s,\ \Delta s > l \qquad \text{(Eq 3.4.1.11)}$$

$$E_2(\theta) = E\times\Delta s,\ \Delta s < l \qquad \text{(Eq 3.4.1.12)}$$

where E= Young's Modulus of Elasticity

Final equation of motion:

$$I\ddot{\theta} + D(\dot{\theta},\theta)\times r_P + (T_1(\theta) + E_1(\theta) - T_2(\theta) - E_1(\theta))\times r_P = F(t)\times KG + mg\times KG\times\sin(\theta)$$
(Eq 3.4.1.13)

The result, a 2nd order with four non-linarites differential equation with components that cannot be algebraically isolated requiring numerical iteration. The solution to a response from a step response of the system is shown in Figure 3.4.1.2 without damping. Figure 3.4.1.3 is the same response with to a system with extreme simulated damping. The parameters used are as follows:

m = 3566321.741; % [kg]

Is=2042.408*(0.3048^4)*(2*m/79.515); %Inertia Tensor x mass/volume, mass moment of inertia [m^4]

cg=5.7912; % [ft] Distance Keel to Center of gravity

g=9.806; % [m/s^2] gravity constant

rad=11.7967; % [m] radius from pivot (rp)

d0=8.1772; % [m] initial distance

l=8.3; % [m] fixed length of chain

rhoAg=127; % [kg/m] mass per unit length of chain

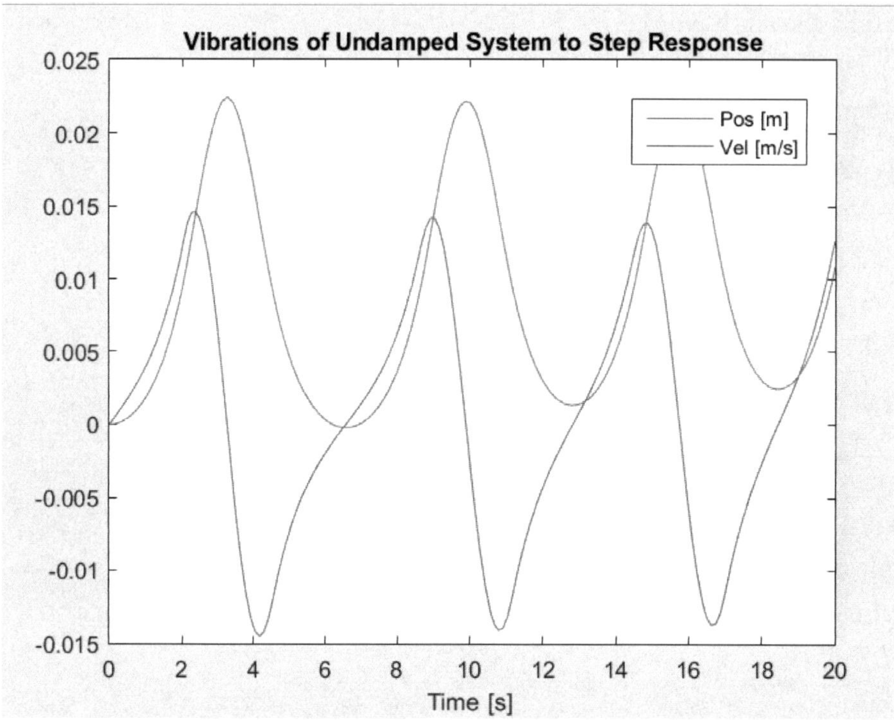

Figure 3.4.1.2: Graph of undamped system given step response of 0.2g

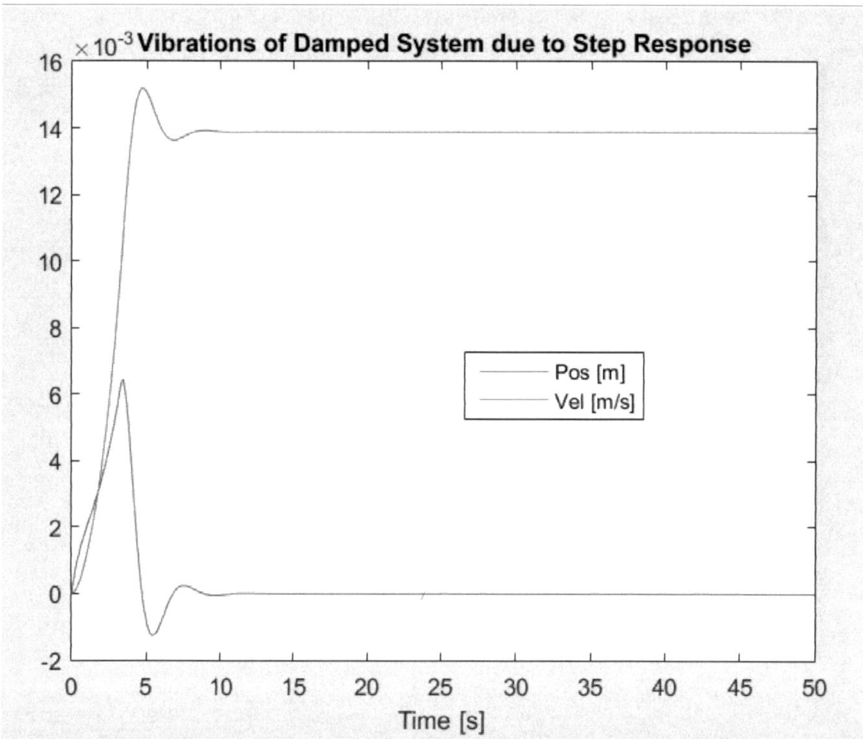

Figure 3.4.1.3: Graph of damped system given step response of 0.2g

The solution from Figure 3.4.1.4 has one of two issues. Observe that the curve is not as we see in 3.4.1.2. The issue here is one of two issues. Either the known issues with the code, or the stability of the system. The stability was not analyzed, and further analysis should be conducted into that specific issue.

The issue with the code is as follows. This was solved using MatLab ODE45 and VPAsolve numerical solver. The ODE45 utilizes the Runge-Kutta method for numerical integration which is optimized for efficiency ODE 45 efficient. However, when this works in tandem with the numerical solver, the interactions cause problems, where iterations in iterations may overlap, makes the analysis unstable. This code would be better optimized with a numerical integration method customized to be compatible with the numerical solver. One of the reasons for this is the catenary's shape, the catenary equation increases in tension at a very high rate right towards the end the chain lengths, essentially where the chain reaches its full extension. The iterations will sometimes shoot right past this point. This will result in the system crashing; other weird results occur as seen in the figure showing instability of iteration system.

Theoretically the system should not have a resonance frequency due to the non-linearity of the system, but due to the instability issues experienced, this will require further analysis and testing.

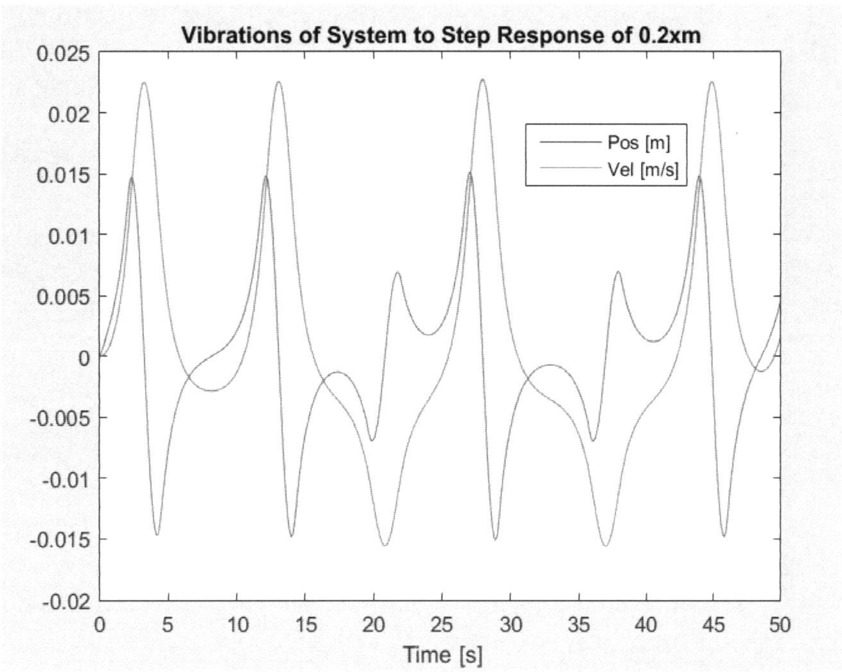

Figure 3.4.1.4: Graph showing of step response of the system showing instability

3.5 Thermal Expansion and Creep Analysis

3.5.1 Purpose

The effects of thermal expansion and creep need to be accounted for to determine whether they can potentially cause a system failure. Normally under short term conditions, the time interval and temperature change are too small to have a noticeable effect on such a large system. However, dry docking procedures can require a vessel to be drydocked up to 6 months or longer. The temperature difference from changing seasons may cause thermal expansion or contraction in multiple areas including the ship, the concrete dry dock, the wooden keel blocks, and the chains themselves. The effects of resting a massive ship on a series of wooden blocks also causes concern for creep effects. Therefore, potential compression of the wood over a period of several months needs to be analyzed as well. The ultimate purpose of these analyses will be to determine the maximum deformation of the different components of the system under the most extreme time and temperature differences. From there, the effects on system efficiency can be determined.

3.5.2 Dimensional Approach

There are several components that need to be analyzed, but not all deformations need to be calculated in three dimensions. The ship itself will be analyzed in three dimensions. However, to simplify things, it will be treated as one uniform steel rectangular prism. This block will expand from the center in both the longitudinal and transverse directions (the x and y directions). It will also expand vertically from the bottom upwards since the bottom is fixed at the keel blocks. See Figure

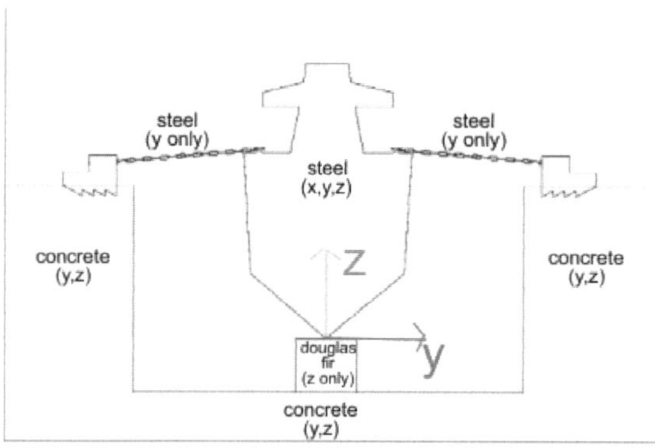

Figure 3.5.2.1: Dimensional Analysis Approach for Thermal Expansion and Creep

The concrete deformation will be analyzed in both the transverse and vertical directions (the y and z directions). This includes both the walls and the floor, which will of course be analyzed only vertically. The keel block will be analyzed for its softwood Douglas Fir texture. Its expansion or contraction is only significant in the vertical direction (z). Last, the steel chains will be treated as a vector rather than a three-dimensional component. Its deformation will only be accounted for transversely so that it will change length-wise.

3.5.3 Thermal Expansion Formula

General expression for change in geometric length due to temperature differences:

$$\Delta L = L_0 \alpha \Delta T \qquad \text{(Eq 3.5.3.1)}$$

where:

ΔL = change in length

L_0 = initial length (or x_0, y_0, and z_0 depending on the direction)

α = thermal expansion coefficient (material property)

ΔT = change in temperature

3.5.4 Thermal Expansion Parameters

The thermal expansion needs to be calculated for extreme conditions to ensure the stability of the system under the most hazardous scenarios. The temperature range used will be from approximately -17.78 °C to 37.78 °C (0 °F to 100 °F). This leaves a ΔT value of about ±55.56 °C or ±100 °F, depending on the units of the thermal expansion coefficient. The sign indicates either thermal expansion or thermal contraction. The distinction is entirely dependent on the transition from hotter to colder temperatures or vice versa.

The ship will be treated as one uniform steel block. The following dimensions and coefficient of thermal expansion will be used:

Longitudinal length (x_0): 152.4 m (500 ft)

Transverse width (y_0): 15.24 m (50 ft)

Vertical height (z_0): 9.144 m (30 ft)

$\alpha_{st} = 6.7\times10^{-6}/^{o}C$

The concrete will be measured transversely and vertically. The following dimensions and coefficient of thermal expansion will be used:

Transverse width/thickness of wall (y_0): 3.048 m (10 ft)

Vertical height/thickness of floor (z_0): 3.048 m (10 ft)

$\alpha_c = 10\times10^{-6}/^{o}C$

The chains will be measured by their change in length. They will be calculated using the same thermal expansion coefficient as that of the ship. The parameters are as follows:

Transverse length (y_0): 6.096 m (20 ft)

$\alpha_{st} = 6.7\times10^{-6}/^{o}C$

The softwood layer. The height and thermal expansion coefficient of Douglas Fir are given below:

Vertical height (z_0): 0.152 m (6 in)

$\alpha_{df} = 27\times10^{-6}/^oC$

3.5.5 Thermal Expansion Results

Deformation in ship:

$\Delta x = \pm 56.718\ mm\ (2.233\ in)$ or $\pm 28.359\ mm\ (1.117\ in)$ from center in both directions

$\Delta y = \pm 5.672\ mm\ (0.223\ in)$ or $\pm 2.836\ (0.112\ in)$ from center in both directions

$\Delta z = \pm 3.404\ mm\ (0.134\ in)$

Deformation in concrete wall and floor:

$\Delta y = \pm 1.702\ mm\ (0.067\ in)$

$\Delta z = \pm 1.702\ mm\ (0.067\ in)$

Deformation in steel chain length:

$\Delta y = \pm 2.261\ mm\ (0.089\ in)$

Deformation in Douglas Fir keel blocks:

$\Delta z = \pm 0.229\ mm\ (0.009\ in)$

3.5.6 Creep Formula

General expression for the rate of change in strain due to creep:

$$\frac{\partial \varepsilon}{\partial t} = A\sigma^n e^{\left(\frac{-Q}{RT}\right)} \qquad \text{(Eq 3.5.6.1)}$$

where:

ε = strain (dimensionless)

t = time interval (in seconds)

A = material constant

σ = applied stress on keel block surfaces from ship load

n = creep exponent (material property)

Q = activation energy (material property)

R = gas constant

T = temperature (in Kelvin)

Rearranging and integrating the formula, the following expression can be used:

$$\varepsilon = \varepsilon_0 + tA\sigma^n e^{\left(\frac{-Q}{RT}\right)} \qquad \text{(Eq 3.5.6.2)}$$

since $t_0 = 0$ s.

In order to solve this equation in terms of length, the strain components are simplified accordingly:

$$\varepsilon = \frac{\Delta L}{L} \quad \text{(Eq 3.5.6.3)}$$

$$\varepsilon_0 = \frac{\sigma}{E} \quad \text{(Eq 3.5.6.4)}$$

since ε will yield the change in length, and ε_o represents the strain due to the initial stress at t = 0 s. Note that E is the elastic modulus, or Young's modulus, of Douglas Fir wood.

Ultimately, the final simplified equation is as follows:

$$\Delta L = L\left[\frac{\sigma}{E} + tA\sigma^n e^{\left(\frac{-Q}{RT}\right)}\right] \quad \text{(Eq 3.5.6.5)}$$

3.5.7 Creep Parameters

The creep in the Douglas Fir also needs to be calculated for the most extreme conditions. Since the chain experiences the most tension under cold conditions, the creep will be calculated at the lowest temperature used in thermal expansion. This corresponds to about -17.778 °C (0 °F).

The activation energy, Q, for Douglas Fir timber that is submerged and extracted from water is about 41.5 $\frac{kJ}{mol}$, or 41,500 $\frac{J}{mol}$.. The gas constant, R, is 8.314 $\frac{J\cdot mol}{K}$. For Douglas Fir undergoing this type of creep mechanism, the constant A and exponent n are 3×10^{10} and 1, respectively. The elastic modulus of Douglas Fir, E, is 12.17 GPa.

The initial height of the keel block, L, is about 0.305 m (6 ft). Assuming a maximum dry docking procedure of 6 months with about 30 days per month, the time period, t, is approximately 15,552,000 seconds. The stress, σ, caused by the ship load of approximately 3.487×10^7 (3,500 LT) resting on about 14.415 m² of keel block surface area, is approximately 2.419 MPa.

3.5.8 Creep Results

Deformation in Douglas Fir keel blocks:

$\boldsymbol{\Delta z = -0.235\ mm\ (-0.017\ in)}$

3.5.9 Overall Dimensional Changes Affecting One Chain (Contraction)

$\Delta x = 28.359\ mm\ (1.117\ in)$

$\Delta y = 20.981\ mm\ (0.826\ in)$

$\Delta z = 5.57\ mm\ (0.227\ in)$

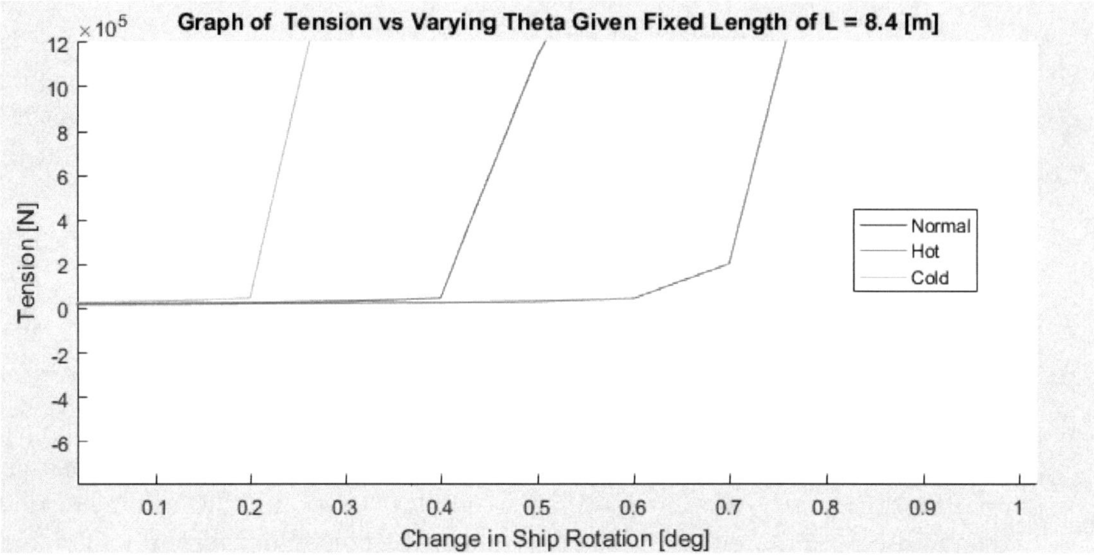

Figure 3.5.9.1: Figure depicting overall dimensional change

As we can see from Figure 3.5.9.1 the maximum contraction under the worst creep and thermal conditions (temperature lowered by 100 degrees F) the catenary curve for tension vs theta there is almost no change in tension. The real effect would happen after approximately 300 degrees F change. Then the curve would intersect the y axis at some much higher tension. If the tension gets too high the chain could snap.

To mitigate this concern, we can measure the chain when it is cold. If the chain gets hotter, there is not a concern as the result is that the ship would just roll farther under the same force.

3.6 Large Scale Component Analysis

3.6.1 Component Identification

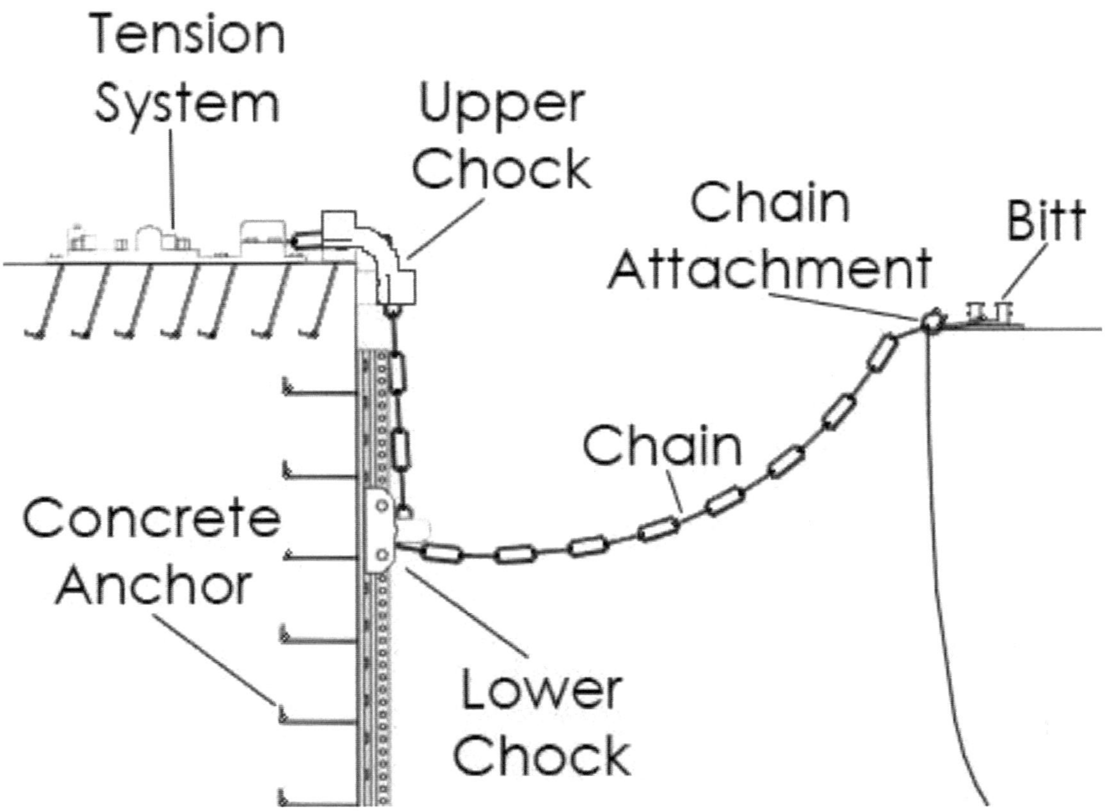

Figure 3.6.2.1: Component Identification Diagram

3.6.2 Ship-side Attachment

The use of a system of chains to maintain stability in dry-docking gives way to multiple components that require appropriate analysis. This analysis focuses on the point of contact between the chains and the ship itself. More specifically, it focuses on how the chains can be best anchored to different parts of the ship.

Chains are not commonly used in the industry of dry-docking ships. In fact, few dry-docks have used them as a means of maintaining ships in a stable environment. There is no available analysis on drydocking with chains. There are certain functional requirements of this component that were taken into consideration:

- The mechanism must be able to allow one end of a chain or chain-like tether to be easily and conveniently attached in a secure manner.
- It must be duplicated along multiple areas on the side of the ship so that multiple points of contact can be made.

- The integrity of the chain is a separate component analysis, but this mechanism must also be able to withstand a sufficient load (approximately 996.4 kN (100LT) per chain given a 5-chain system).

Bitts:

Ship bitts consist of two strong posts fixed vertically into a bottom surface. In this case, it would be fixed into the dock of the ship. The chain wraps around the bitts in multiple figure-eight type loops. Table 3.6.2.2 displays the safe working loads of bitts depending on size.

If the loading on the bits exceeds the loading on one bit, a secondary bitt can be used to share the load. See the diagram, Figure 3.6.2.2.

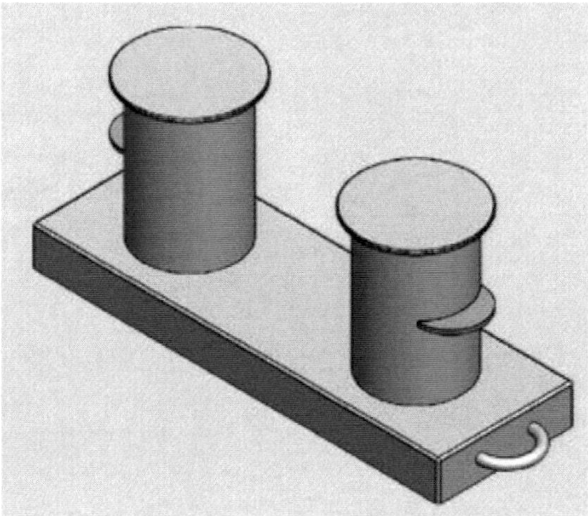

Figure 3.6.2.1: Image of ship bitts

Table 3.6.2.1: Safe Working Load of Steel Bitts (Ref 7)

Bitt Size (Diameter)	SWL (One barrel, any direction)
4 in.	23,000 lbs.
8 in.	60,000 lbs.
10 in.	100,000 lbs.
12 in.	164,000 lbs.
14 in.	265,000 lbs.
18 in.	375,000 lbs.

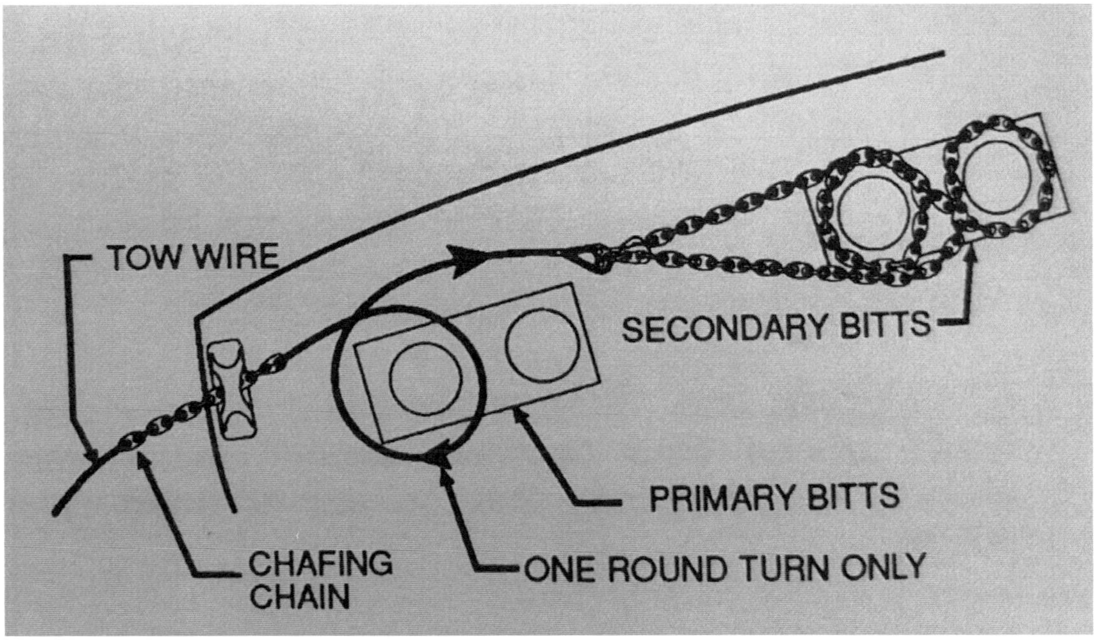

Figure 3.6.2.2: Sharing loads between two bitts (Ref 7)

3.6.3 Ship-side Chain Attachment

The chain will be attached to bits using some components

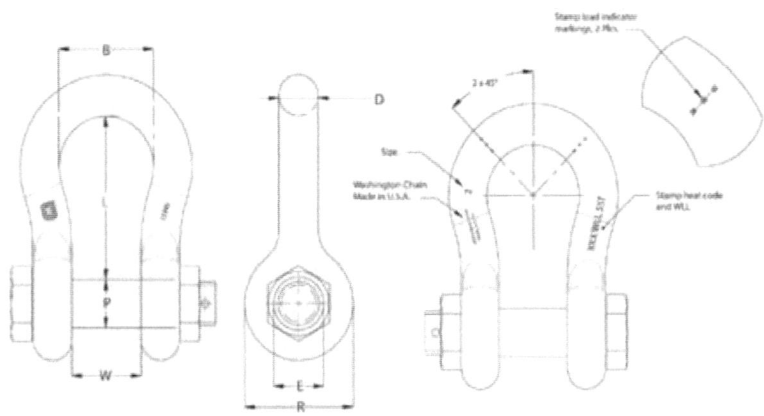

Figure 3.6.3.1: Figure of U-Bolt to connect chain to bitt attachment

Table 3.6.3.1: Table of U-bolt options

Part Number	Trade Size (nom)	D Dimension (min)	P Dimension pin (min)	E Diameter Inside Eye	W Width Between Eyes	B Bow (min)	L Length Inside	R Diameter Outside eye (max)	Working Load Limit (WLL) Tons
8472120003	2	1.9	2.19	2.38+0.020/-0.015	3.250±0.188	5.25	7.75±0.50	5.25	55T
8472125003	2.5	2.38	2.68	2.88+0.020/-0.015	4.125±-0.250	6.75	10.50-±-0.50	6.25	85T
8472130003	3	3.12	3.25	3.38+0.020/-0.015	5.000±0.250	7.375	13.00±0.75	6.75	120T
8472135003	3.5	3.62	3.75	3.88+0.030/-0.025	5.500±0.250	9	15.00±0.75	8.5	150T
8472140003	4	4.0	4.25	4.38+0.030/-0.025	6.500±0.250	10.5	17.00±0.75	9.5	175T

3.6.4 Chain

Functional requirements:

Selecting the optimal chain/cable for stabilization of various types of ships for dry-docking, requires a component analysis to ensure that the risk of failure for the chosen component is minimal. Using a minimum of five chains on each side of the ship for support, the minimum withstanding load for each chain must be 90720kgs (100LT). Since safe working loads and breaking strengths for any grade chain are two-thirds (safety factor of 1.5), the ideal chain must have a minimum breaking strength of 150LT or 1494.6kN. If this condition is not met, the number of chains required must be increased until the load requirement is met.

Additionally, the selected chain must be able to withstand extreme weather conditions, and keep its minimum withstanding load over time. It is also necessary for the chosen chain to be compatible with several different types of ship anchors and mechanisms.

Stud-Link Chain

Diameter: Up to 0.0254 meters (4 inches) – Loading capacity: 518,000kgs (1,143,000lb) – Elongation at Break: <1%

Characteristics: Stud-Link chains are stronger and more kink-resistant than other types of chains and cables, suitable for use with small bending radius, will not elongate over time

Figure 3.6.4.1: Photograph of stud-link chain

Current Applications: Towing bridles, connecting to attachment points on ships, towing.

Satisfying conditions of Stud-Link Chains: Chains were chosen over other components options described in table because of their loading capabilities and durability. The stud-link chains sufficient load capacity to withstand the rotational movement of the ship due to any applied force (a level 7 earthquake in our experiment). Additionally, chains are already used at a docking site in Rotterdam, Netherlands and throughout the shipyard industry. A table with varying chain link dimensions and loading capacities can be found in Appendix. Specifications for other component considerations can also be found in the Appendix.

Table 3.6.4.1: Table comparing chain component options

Component	Pros	Cons
Kevlar 12-Strand Fiber Ropes	-High stiffness), toughness, and thermal stability -Will not rust over time -No electrical conductivity -Will not become brittle at lower temperatures -Often used for towing ships	-Weaker than wire rope and stud-link chain, and would require a minimum of 8 ropes on each side -Not as heavy as metals, and would sway under heavy winds -Absorbs moisture
Wire Rope	-Unlike chains, strength is dominated by tensile strength, so single point failure less likely to cause total failure under tension	-Elongates over time -Difficult to attach to ship without additional mechanisms -Not known to be used for applications near water
Stud-Link Chain	-Very high minimum breaking strength -Less likely than other chains to deform when overloaded -Compatible with many different types of ship anchoring mechanisms -Less expensive than wire rope and 12-Strand Kevlar	-Links can turn under rotational forces -Single point failure would cause complete failure

3.6.5 Routing System

The routing system is designed to change the height of the dock side attachment. It consists of the upper chock and lower chock. The upper chock is a somewhat standard chock. The lower chock is adjusted vertically in position to allow the chain to reach the best righting angle to the ship when under tension. The lower chock would have to be custom made and need the ability to be locked into the rail at the specified height via a pin. This would allow for lower tension in the chains due to a more direct righting moment. Refer back to Figure 3.6.2.1 to view the upper and lower chocks.

3.6.6 Tension System

Functional requirements:

One of the major components of the modified system for docking ships is a tension system, which is essential for keeping the slack to a minimum. The chains must be as taut as possible to reduce the possibility of any movement. Some mechanisms that serve as possible options are grip hoists, power pull manual chain hoist, or an elevator motor system.

Component options:

Figure 3.6.6.1: Drawing depicting operation of a griphoist

Grip hoist

The grip hoist consists of two pairs of jaws that alternate pulling on the wire being tightened to avoid losing any tension. These jaws are locked by the pull of the wire rope. Therefore, the heavier the load is, the stronger the grip.

Power pull manual chain hoist

The power pull manual chain hoist is like the grip hoist but uses a set of chains and multiple gears to transfer a small force with high travel input on one chain into the low movement, high force output on another chain. This system essentially uses mechanical advantage to lift the load. This system has a ratchet that prevents the load from back driving the input. The ratchet is disengaged by a clutch when back driving is desired and applied on the smaller chain.

Elevator system

The final system that could work as the tensioning system, the elevator system, is made up of a motor, pulleys, a counterweight and a braking system. The motor is mainly used to determine the direction of motion, while the counterweight is used to balance the system. The motors used on elevators have a governor that contains spring loaded strings called flyweights that serve as emergency locks in case the elevator starts moving too quickly. With some modifications this system can be incorporated into the drydocking industry.

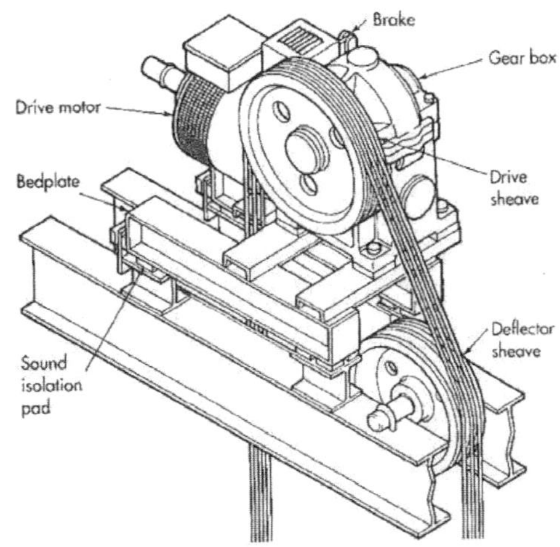

Figure 3.6.6.2: Diagram depicting elevator hoisting system

Table 3.6.6.1: Table comparing tension system options

	Pros	Cons
Grip hoist	-Cost: $1,300-4,000	-No fail-safe mechanism -Manual system -Working load limit(WLL): Pull: 7257kg (16,000lb) Lift: 3629kg (8,000lb)
Pull manual Chain hoist	-Cost: $160	-No fail safety mechanism -Manual system -Working load limit (WLL): 2722kgs (3 Tons)
Elevator system	-Fail safety mechanism -Automated system -Working load limit: Counterweight	-Complex system -Limited spec availability

3.6.7 Spring/Damper System

The spring/damper provides damping to the system. Without it, the system could become excited under various earthquake or hurricane loads.

"Big Ton" (shown Figure 3.6.7.1) multi-spring design lends itself to supporting very high loads. The load values in the table may easily be modified to accommodate the client's needs. Loads up to 200,000 pounds can be handled with no problems. The designs which appear in the load table are "off-the-shelf" designs and can be shipped with good delivery dates. Designs above those shown in the load table will be furnished upon request.

Figure 3.6.7.1: Heavy Load "Big Ton" Spring Supports for a Chemical Processing Plant

Methods of damping are viscous damping, or something with internal damping like a leaf spring. See Figure 3.6.7.2.

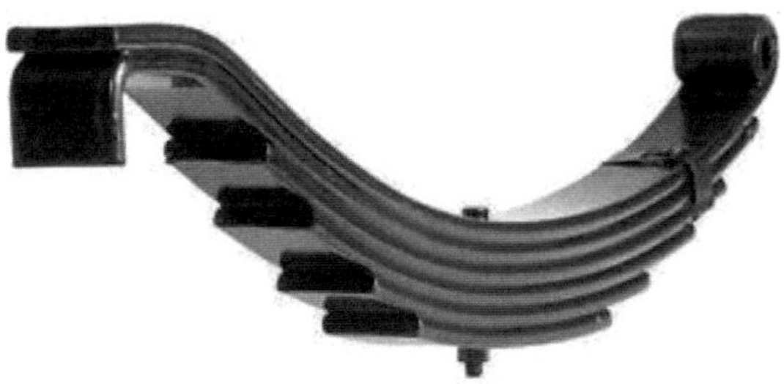

Figure 3.6.7.2: Image Depicting a Leaf Spring

3.6.8 Concrete anchor

Functional requirements:

One requirement of the method is keeping a chain in tension imbedded in a concrete wall or floor. It must withstand a load of 90720kg (100LT). Given the approximation of 453,592kg (500LT) per side and use of five chains per side. Working load safety factor of 4:1out lined by ASTM standards. ASTM International is an international standards organization that develops and publishes voluntary consensus technical standards for a wide range of materials, products, systems, and services.

Bonded cast-in-place anchors

These anchors are locked in place by bonding, but require excavation of the area to be casted in place.
Types: Stress gradient, embedded plate-nut, J-bolts, L-bolts, bent rebar

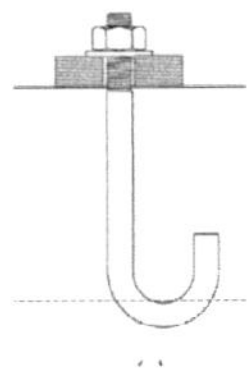

Figure 3.6.8.1: Diagram depicting a J-bolt concrete anchor

Satisfying conditions of the casted anchor: This form of anchoring is the best option for us since we are working with loads that are in the tons and these anchors provide the highest degree of dependability. Multiple anchors can be used to distribute the tensile load applied on the tensioning system. We used L-bolts to secure our tensioning system. Additional considered component options can be found in the Appendix.

Table 3.6.8.1: Table comparing concrete anchor options

	Mechanical Anchors	Casted Anchors	Epoxy Anchors
Pros	- A pre-stressed anchor prevents anchor fatigue failure in cyclic or dynamic loading applications - Anchors can be custom manufactured or mechanically coupled to any length	- Strongest anchors and high degree of dependability - Anchors can be custom manufactured or mechanically coupled to any length	- Capable of reaching high bond-stress values in relatively fast cure times
Cons	- Lower strength	- Require more severe modification of the area and no ability of replacement without significant modification	- No longer used in construction - Are prone of failure under freeze/thaw conditions

3.7 FMECA

Failure mode, effects, and criticality analysis (FMECA) was conducted on the proposed chains system to determine the critical components and their corresponding failure concerns. (See Appendix) The chains system was divided into four major components: the ship side attachment, the chain, the routing system, and the dock side attachment (Table 3.7.1). The purpose of the analysis was to determine which of these components was the most concerning when performing a docking. To determine this, each component was divided into its subcomponents. After outlining each subcomponent's functionality, possible failure modes for each subcomponent was considered and it was concluded that most failures would originate from either material fractures or corrosion. The most concerning failure was a rupture in the actual chains. If multiple chain failures occurred, this could result in damage to the vessel or to the crew working on the ship. The criticality for all the failure modes was based on the severity of the failure (50%), the possibility of occurrence (15%), the detectability (30%), and the difficulty in fixing the issue (5%). These four categories received a value between 1 and 5; one being the most critical and five being the least critical. The assigned values were then proportionally summed up to determine the criticality value. For example, the chain component received a 1 in severity, a 4 for occurrence, a 1 for detection and a 5 for difficulty fixing. These values were multiplied by their corresponding percentages and then added together to determine the criticality value of 1.65. The severity of the failure carried the greatest percentage due to the fact that it implied that the failure mode occurred. This would call for a course of action to minimize damages. The next category that carried that greatest percentage was the detectability since this could prevent any failures. The easier a failure could be detected; the sooner it could be prevented. Occurrence received a value of 15 percent because it considered the number of each subcomponent. The more of each subcomponent; the greater chance of failure. Finally, difficulty fixing the subcomponent had the lowest impact on the criticality because this is a post failure consideration.

In addition to the criticality value, the FMECA contained corrective actions to take at different stages of the operation. In the case when the operation had begun and a failure occurred, an emergency course of action could be taken to proceed with the docking operation. Most of these actions required a re-routing of the chain through either the upper chock or through an adjacent tensioning component. These changes would result in minimal effects outlined in the analysis. In addition to the emergency scenario, other corrective actions were outlined for failures that were detected before the docking operation started. The corrective actions were outlined for both minor and major failures. Finally, precautionary courses of action were described that would add redundancy to any given subcomponent.

Table 3.7.1: Summary of the most to least critical components

Component	Criticality
Chain	1.65
Dock side attachment	2.7
Ship side attachment	3.6
Routing system	4.1

3.8 Cost Analysis

Cost analysis was done to compare the proposed docking system (chains) to that of the current system (side blocks). Before performing this analysis, it was estimated that the system of chains would require a higher startup cost, but that it would eventually yield greater profits due to the minimization of waste. Quotes from a ship supplier were acquired that carried most of the components used in the chain system. All values received were based on the expected working lifetime of any given component. For example, the standard blocks used for the keel of the ship cost $3,000 each and have an estimated life span of 25 years. To perform the cost analysis, three separate sections were developed to accurately account for all the different sources of cost: startup, material/equipment, and labor. The startup cost section consisted of purchased materials that had a lifespan greater or equal to a year. The material/equipment section outlined all the components that had a lifespan of a single use. This meant that they were used for a single docking and then discarded or returned to the lessor. The final section of the analysis, labor, considered the skilled and unskilled work hours required for the preparation and operation of the docking process. These three sections were integrated to acquire the yearly cost of operating the dry dock. The number of dry docks performed per year varied based on whether a naval or commercial vessel was being docked. In addition, it was estimated that the chain docking system would speed up the docking process from 3 weeks for naval ships and 2 weeks for commercial ships to one week for either.

The yearly cumulative cost was calculated for a span of 25 years. The cost of the startup materials was reintroduced to the corresponding yearly cost once the previous component lifetime expired. The chain system required less preparation time to dry dock which allowed for 10 naval docking and 24 commercial dockings a year in comparison to the 7 and 16 for side block drydockings. The profit for the side block system was determined based on a percentage markup of the total cost per dry docking. The profit for chains was based on a similar percentage markup plus the difference in cost of chains and side blocks. This basically assumes that the cost the yard is pricing the services equally. This is a good approximation because the result was both systems had similar times to profitability.

Over all the chain system shows promise of higher profitability. It has a higher start-up cost (for a renovated dry dock), although it has lower operational costs and can support more drydockings in any given year. After further development, this system should also allow for lower operational risk which would make this system even more attractive. See Table 3.8.1 for a table showing the profit over the lifetime of the system. While Figures 3.8.1, 3.8.2, and 3.8.3 display this data as graphs.

Table 3.8.1: Profit of Side Blocks vs Chain System Drydocking Commercial and Navy ships and the Average

Commercial			Navy			Average		
Time (years)	Profit (SB)	Profit (Chains)	Time (years)	Profit (SB)	Profit (Chains)	Time (years)	Profit (SB)	Profit (Chains)
0	($459,000)	($1,575,200)	0	($459,000)	($1,575,200)	0	($459,000)	($1,575,200)
1	($295,479)	($1,123,765)	1	($241,236)	($995,085)	1	($295,479)	($1,123,765)
2	($131,958)	($672,331)	2	($23,472)	($414,970)	2	($131,958)	($672,331)
3	$31,563	($220,896)	3	$194,292	$165,145	3	$31,563	($220,896)
4	$195,084	$230,538	4	$412,056	$745,260	4	$195,084	$230,538
5	$358,605	$681,973	5	$629,820	$1,325,375	5	$358,605	$681,973
6	$522,126	$1,133,407	6	$847,584	$1,905,490	6	$522,126	$1,133,407
7	$685,647	$1,584,842	7	$1,065,348	$2,485,605	7	$685,647	$1,584,842
8	$849,168	$2,036,276	8	$1,283,112	$3,065,720	8	$849,168	$2,036,276
9	$1,012,689	$2,487,711	9	$1,500,876	$3,645,835	9	$1,012,689	$2,487,711
10	$1,176,210	$2,939,145	10	$1,718,640	$4,225,950	10	$1,176,210	$2,939,145
11	$1,339,731	$3,390,580	11	$1,936,404	$4,806,065	11	$1,339,731	$3,390,580
12	$1,503,252	$3,842,014	12	$2,154,168	$5,386,180	12	$1,503,252	$3,842,014
13	$1,666,773	$4,293,449	13	$2,371,932	$5,966,295	13	$1,666,773	$4,293,449
14	$1,830,294	$4,744,883	14	$2,589,696	$6,546,410	14	$1,830,294	$4,744,883
15	$1,993,815	$5,196,318	15	$2,807,460	$7,126,525	15	$1,993,815	$5,196,318
16	$2,157,336	$5,647,752	16	$3,025,224	$7,706,640	16	$2,157,336	$5,647,752
17	$2,320,857	$6,099,187	17	$3,242,988	$8,286,755	17	$2,320,857	$6,099,187
18	$2,484,378	$6,550,621	18	$3,460,752	$8,866,871	18	$2,484,378	$6,550,621
19	$2,647,899	$7,002,056	19	$3,678,516	$9,446,986	19	$2,647,899	$7,002,056
20	$2,811,420	$7,453,490	20	$3,896,280	$10,027,101	20	$2,811,420	$7,453,490
21	$2,974,941	$7,904,925	21	$4,114,044	$10,607,216	21	$2,974,941	$7,904,925
22	$3,138,462	$8,356,359	22	$4,331,808	$11,187,331	22	$3,138,462	$8,356,359
23	$3,301,983	$8,807,794	23	$4,549,572	$11,767,446	23	$3,301,983	$8,807,794
24	$3,465,504	$9,259,228	24	$4,767,336	$12,347,561	24	$3,465,504	$9,259,228

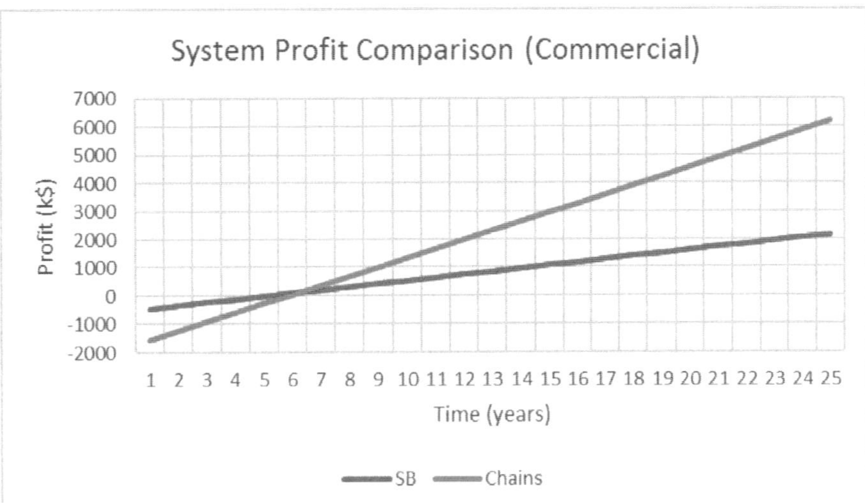

Figure 3.8.1: Profit comparison of docking commercial vessels with Side Blocks (SB) and the Chains system over a span of 25 years drydocking comeritial ships

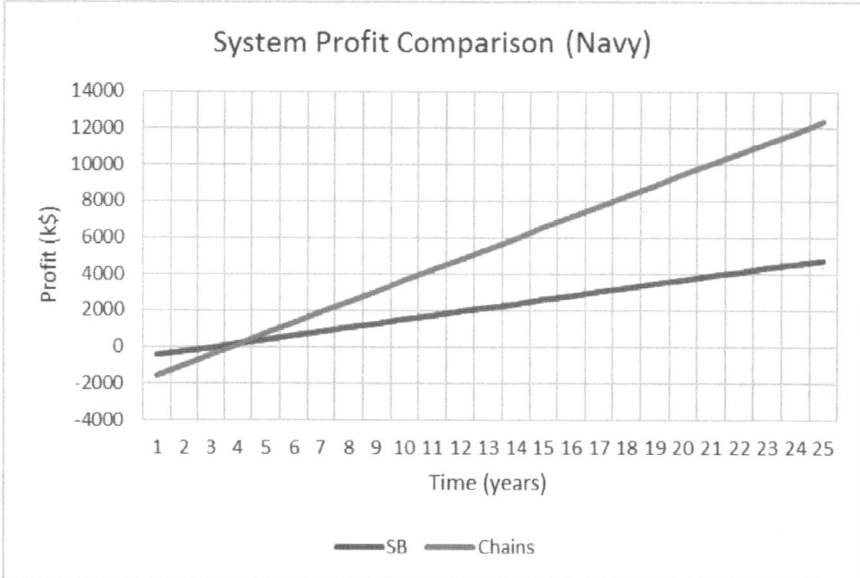

Figure 3.8.2: Profit comparison of docking navy vessels with Side Blocks (SB) and the Chains system over a span of 25 years.

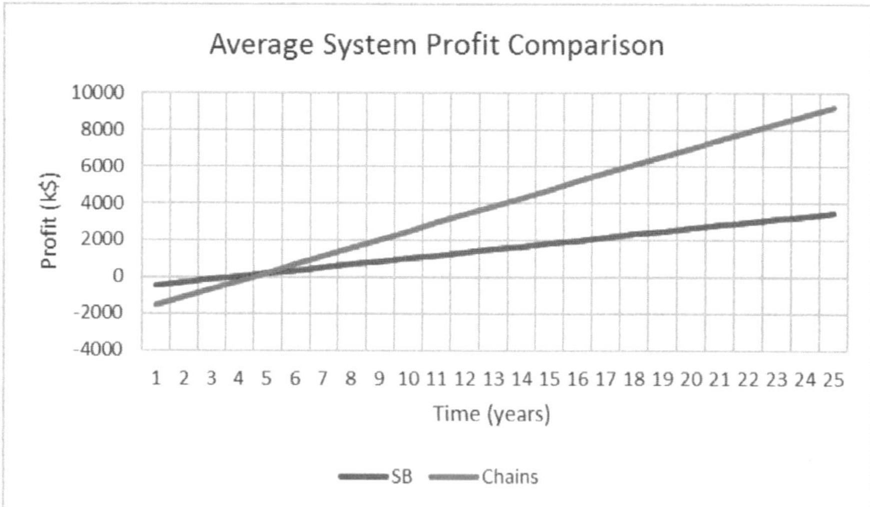

Figure 3.8.3: Average profit comparison of docking vessels with Side Blocks (SB) and the Chains system over a span of 25 years.

Full cost analysis in the Appendix.

Chapter 4: Future Considerations

4.1 Design recommendations for the future

Seismic Derivation Analysis

The seismic analysis derivation may need more consideration. The overturning forces that arise from earthquake or hurricane are derived from Ref 1. But these derivations which is intended for a side block system. The force may not actually act at the center of gravity of the ship. It would make sense that the earthquake force would be the force of friction causing the ship to move. We assumed that the overturning force should be the same for both systems early on, although after consideration, it may not be the case. This would change some of the static and dynamic analysis, so careful examination of the derivation for the seismic and hurricane should be made. Also for the dynamic analysis, the force rotates with the ship in an overturning moment, and it may be more accurate to keep the force parallel with the earth.

Slip

The dynamic and static analysis showed different results for slip. The issue was that in an upright position $\theta=0$ [deg] the catenary of each equation cancels horizontal components. Therefore, the sum of all horizontal components are the overturning force and friction. So, $F=f$, or $ma=uN$. The tension in the chains in the upright position is minimized leaving a very small vertical component that would attribute to N. So, by approximation $N=mg$, and $a= 0.2g$ we find the coefficient of friction $u=0.2$ By looking at results in Ref 8 we see that depending on the moister level of the wood the coefficient of friction is 0.15-0.25. So potentially the slip could be an issue for a 0.2g earthquake. One factor that is not calculated in this assessment is that the keel will sink into the soft cap reducing the ability to slip. This risk can also be mitigated by adding rubber caps on the blocks which would increase the coefficient of friction. Alternatively, additional wedge supports could be added in after the dry dock has been pumped dry to prevent the keel from slipping off. This would reduce the risk to only the short period of time before extra supports can be installed. Note, one would not want to install extra keel supports before landing the vessel because the ship could land on these supports causing a point load on the hull. It is normal dry docking procedure to avoid these point loads and have 0.3[m] (12 [in]) of clearance between a smooth landing area and any obstruction. Longitudinal slip could also be a concern for the same reasons.

Catenary Jerking Force

The catenary analysis showed a very steep non-linear trend for tension as the sag in the rope dropped to zero. This non-linear trend is advantageous for minimizing the tension in the chain while the ship is not experiencing large forces, like an earthquake or hurricane. However, due to a huge spike in tension right before the chain becomes taught, there would be a large jerk force (large change in acceleration). It should be noted that this force would be finite and continuous with time, but should still be considered. To minimize jerk, the steps for optimizing the catenary non-linearity should be reversed. See catenary section in chapter 3 for details. Therefore, seek the longest distance between chain attachments, and the largest height differential between chain attachments.

Irregular Loading Across Ship

Due to efficient hydrodynamic shape of the ship, the distance between the chain attachment on the ship and nearest point on the dock is not uniform at all points longitudinally. This is potentially an issue with a system relying on catenary non-linear tension to right the ship. While each chain can be optimized to exhibit the same righting moment at a certain degree of ship movement, each chain will not all apply the same loading for each degree on its path to the full righting moment. The issue is that the loading will be higher at some points longitudinally on the ship and lower at others. This could potentially cause some issues, especially with the jerking response at mentioned above.

Concrete Reinforcement

The concrete anchor research was somewhat limited in span due to other requirements of the project. The concrete analysis done proved that the concrete would be capable of withstand the loading. More time could be spent analyzing the structure of the dry dock and the specific loading capabilities.

Chain Sleeves

The option of chain sleeves that would fit over the chains was briefly considered. This could provide corrosion protection, allow for smoother passage over the fittings to protect the dry dock and the ship, and provide additional support in the event of the chain failure. If the chain were to fail, the chain could fall into the basin of the dry dock and cause further damage to the dry dock and ship or cause injury to personal. Alternatively, a wire could be inserted in the center of the chain to provide the support in event of chain failure.

Dynamic Consideration

The dynamic system of equations was all developed for the sum of the forces and the moments is equal to zero. While this is the case for the static analysis, this should not be the case for the dynamic in most cases. The formulas could have been sum of F = ma where a = some acceleration of the body. This would have changed the analysis and results.

Chock Bending Radius

The bending radius of the chains needs to be at least 7 times the bar that was used to create the chain (the internal diameter of the chain) (Ref 7). The surface of the chocks need to be smooth such that

Maximum Angles and Loading

The maximum angles in each of the three axes, and the loading on the chains needs to be established for each dry dock.

Stability

A calculation for stability of the system should be completed. The analysis done seem to indicate that there will be stability, however a more in-depth stability calculation should be performed.

Vibrational Chain Damping

The vibrations chain damping energy balance is uses a viscus damping approximation. Viscous damping is a function of $\dot{\theta}$ but the system could better be modeled as frictional damping which is a function of θ. Both are only small contributions to damping. Verification of the direct contribution to the system should be made. The coefficient of friction between the chain links should also be verified.

4.2 Final Considerations

Safety Considerations

The nature of dry docking make it a dangerous business. The objective of dry docking is to support ships out of water with minimal interference for repairs and maintenance. Ships are heavy, and put large loads on the dry docks and supports. Mishandling of a ship can and has caused damage to the ship, the dry dock, and injury to personnel. Many dry dock yards are pressured to dry dock ships quickly and efficiently to turn a profit. The increase safety can save ships and dry docks from damage, and save personnel from injury.

Environmental Impact

Our system would greatly reduce the waste produced by the current system of dry docking. Dry docking with side blocks requires individual side blocks to be cut to the shape of the ship and placed precisely where they can support the ship. The blocks are divided into 3 parts, the base block, the hard wood section and the softwood section. The base block is a cement block that can be used for many years without much maintenance. The hard wood section is generally reused for several operations, although can become unusable from being soaked in salt water and then dried in the sun. The softwood portion of the block is cut to the shape of the ship, and is normally discarded after a few operations. Soft-wood is usually used for a few dry dockings.

Acknowledgements

Special thanks the following people for their assistance in this project:

Jonathan Leyva, BS Mechanical Engineering, UCSD, for his contributions of the entire cost and along with various other contributions and assistance along the way.

Adam Johnson, BS Mechanical Engineering, UCSD, for his hard work on creep and thermal expansion analysis, the write up and various other contributions to the project.

Joe Stiglich, DM Consulting, for sponsoring this project and providing a memorable opportunity to work on a large-scale real-world industry project.

Frank Russel, DM Consulting, and Tom Hengel, Mare Island, San Francisco, for providing industry knowledge to the project.

Tom Chalfant and Ian Richardson in the UCSD Mechanical Enigneering Machine Shop for guiding us in the machine shop and with our production methods.

Jessica Tuazon and Steve in the UCSD SME department for providing a shake table to perform the vibrations test on the 1:50 scale model.

Larry Duby from NASSCO in colaboration with IDNA (Internal Design and Naval Architecture) for providing us with insitefull into ship building expertise.

The folks at Damen Maaskant Shipyards Stellendam in Rotterdam for providing us with helpful insite to a chain system.

Dr. Jerry Tustaniwskyj and Pedro Franco for instructing the UCSD MAE 156B Winter 2017 class that made this possible while helping us along the way.

Dr. Nate Delson for instructing the UCSD MAE 156A Fall 2017 class.

The rest of the UCSD MAE 156A Fall 2016 class and UCSD MAE 156B Winter 2017 class for thier crusial feedback and contriubutions to the project.

References

Reference 1: Industry Drydocking Calculations

MIL-STD 1625

NSTM (Naval Ships Technical Manuals) 997 (Restricted)

DM Consulting, Inc. Art and Science of Drydocking. 2017. Print

Reference 2: Docking Plan for the FFG-7

FFG-7 Docking Plan: 8011 107

NavShips Drawing Number: PF '09 801 46702

Available at: http://www.drydocktraining.com/drawings.html

Reference 3: San Diego Graving Dock Information Sheet

U.S. Naval Station San Diego CA. Graving Dock No. 1 information sheet

See Appendix

Reference 4: Hydrostatic Data Diagram for the FFG-7

D&O curves for the FFG-7

See Appendix

Reference 5: Material Properties

CES Selector – Material Universe

The standard tool for materials selection and graphical analysis of materials properties

Reference 6: Douglas Fir Damping Ratio

Moore, J. R., and D. A. Maguire. "Simulating the Dynamic Behavior of Douglas-fir Trees under Applied Loads by the Finite Element Method." Tree Physiology 28.1 (2008): 75-83. Web.

Reference 7: Miscellaneous Nautical Loading Information

U.S. Navy Salvor's Manual. NavSea S0300-A7-HBK-010

U.S. Navy Salvage Engineer's Handbook, Volume 1: Salvage Engineering. Bartholomew, Marsh, Hooper. 1992. Print.

Reference 8: Wood to Metal Friction Coefficient

Murase, Yasuhide. Friction of Wood Sliding on Various Materials. Thesis. Laboratory of Wood Technology, Faculty of Agriculture, Kyushu University 4G-08, Fukuoka 812, 1983.

Reference 9: Catenary Equations

About the Curve of a Free Hanging Rope. Ruud v Gessel, April, 2001

Reference 10: Thermal Expansion and Creep

Hertzberg, Richard W. Deformation and Fracture Mechanics of Engineering Materials. N.p.: n.p., 1976. Print.

Reference 11: Douglas Fir Creep Mechanisms

Hertzberg, Richard W. Deformation and Fracture Mechanics of Engineering Materials. N.p.: n.p., 1976. Print.

Standards

The navy outlines standard dry dock calculations in Mil Std 1625. The US navy also abides by a more strict set of necessary calculations as outlined by NSTM (Naval Ships' Technical Manuals) 997 which is a restricted document. Mil-std 1625 provides the basis for dry dock calculations utilized around the world. Some of the calculations Mil-Std 1625 outlines that we used in our analysis is: block loading calculations, survival of earthquake forces, survival of hurricane forces, and the draft at instability (the draft of the ship as it land where it becomes unstable).

There is no organization that provides standards to the dry dock industry. Each dry dock yards can set their own set of standards and enforces them as they please. This includes operating procedures, maintenance routines, structural requirements, and personnel experience requirements. Some floating dry docks (being vessels themselves) get certified under large maritime certifications as a barge or a heavy lift vessel, although again this is not a requirement. It is possible that the ASTM building codes apply to construction of graving docks, and pier-side construction codes apply to other dry dock constructions, although it is widely unenforced.

We have also tried to implement other standards into our analysis. We have used building code standards such as ASTM A722 for approved minimum strength of a concrete anchor. As we implement for analysis throughout the quarter we will update this list for a comprehensive set of standards.

Additional website references for images

http://www.christinedemerchant.com/

http://www.barzilay.nl/

http://www.wilderness.net/toolboxes/documents/tools/rigging/grip_hoist_manual.pdf

https://www.youtube.com/watch?v=3j55IAg3GgE

https://www.youtube.com/watch?v=CeOklEyUw0I

https://www.astm.org/Standards/A722.htm

https://law.resource.org/pub/us/cfr/ibr/001/aci.318.1995.pdf

http://www.post-tensioning.org/

http://www.dot.ca.gov/hq/esc/techpubs/manual/bridgemanuals/bridge-design-aids/page/bda_5_81.pdf

http://www.williamsform.com/Contact_Us/PDFs/High_Capacity_Concrete_Anchor_Systems.pdf

Appendix

Appendix A: Additional References

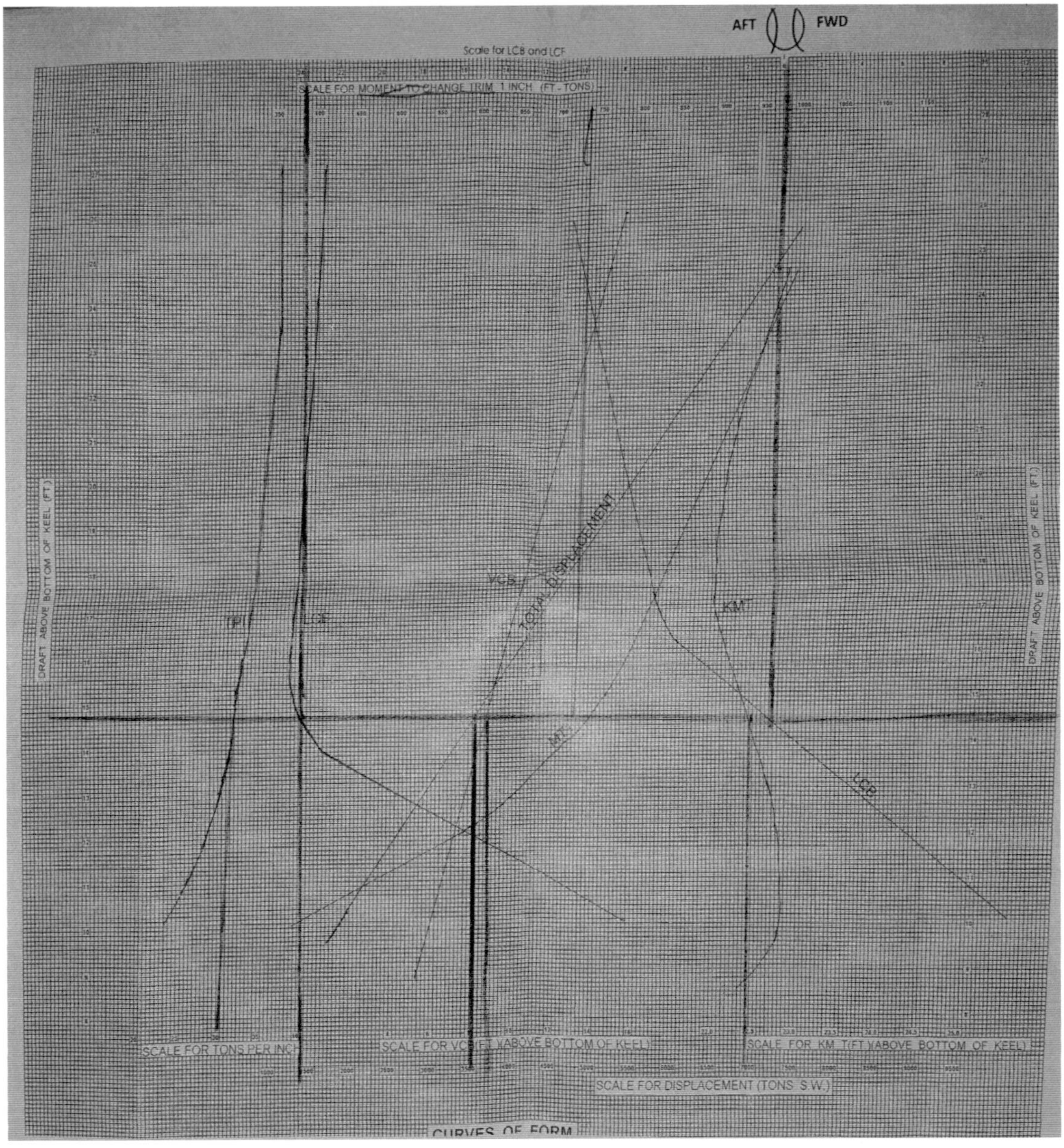

Figure A.1: D&O curves (hydrostatic data) for the FFG-7

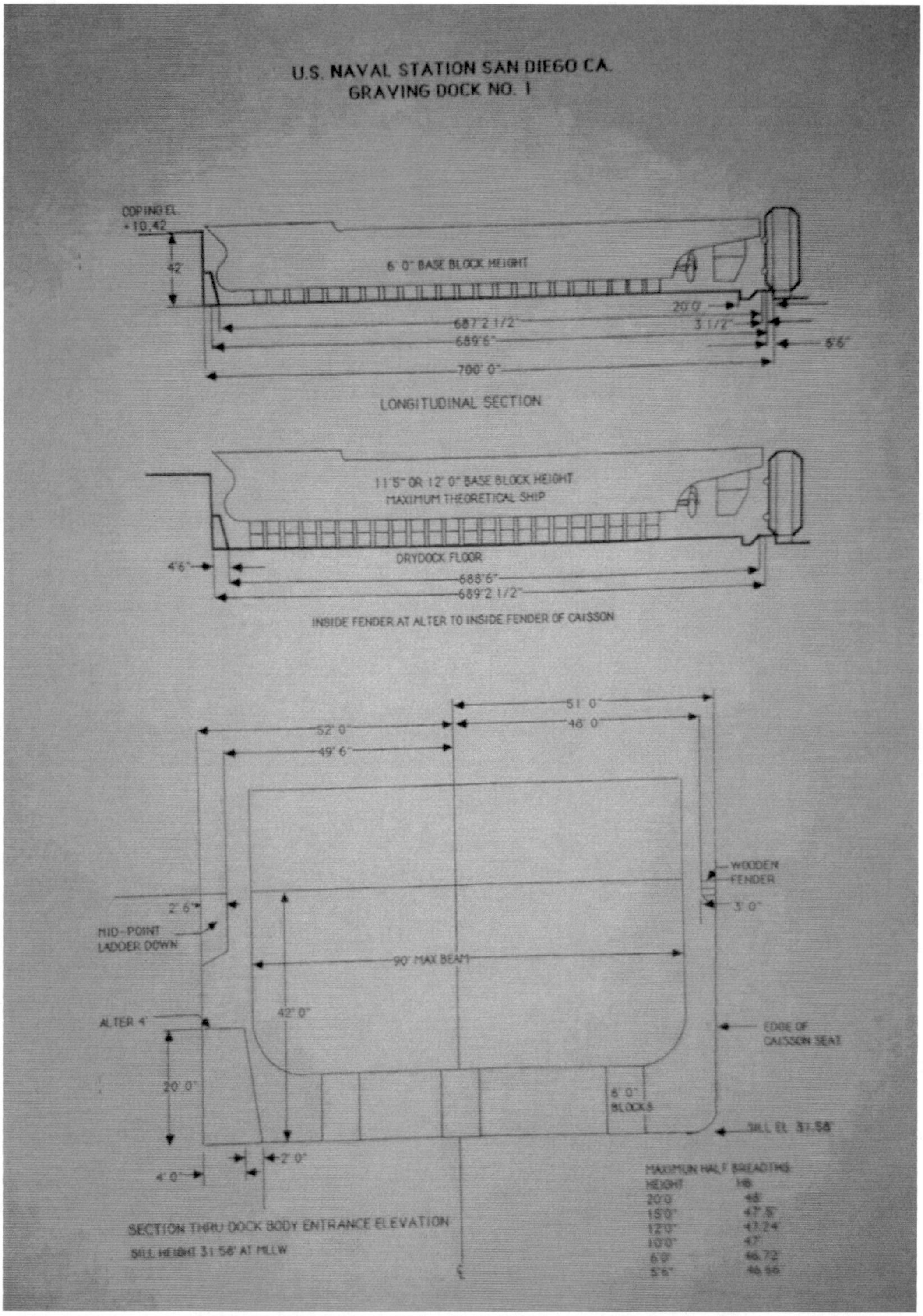

Figure A.2: San Diego Graving Dock Information Sheet page 1/2

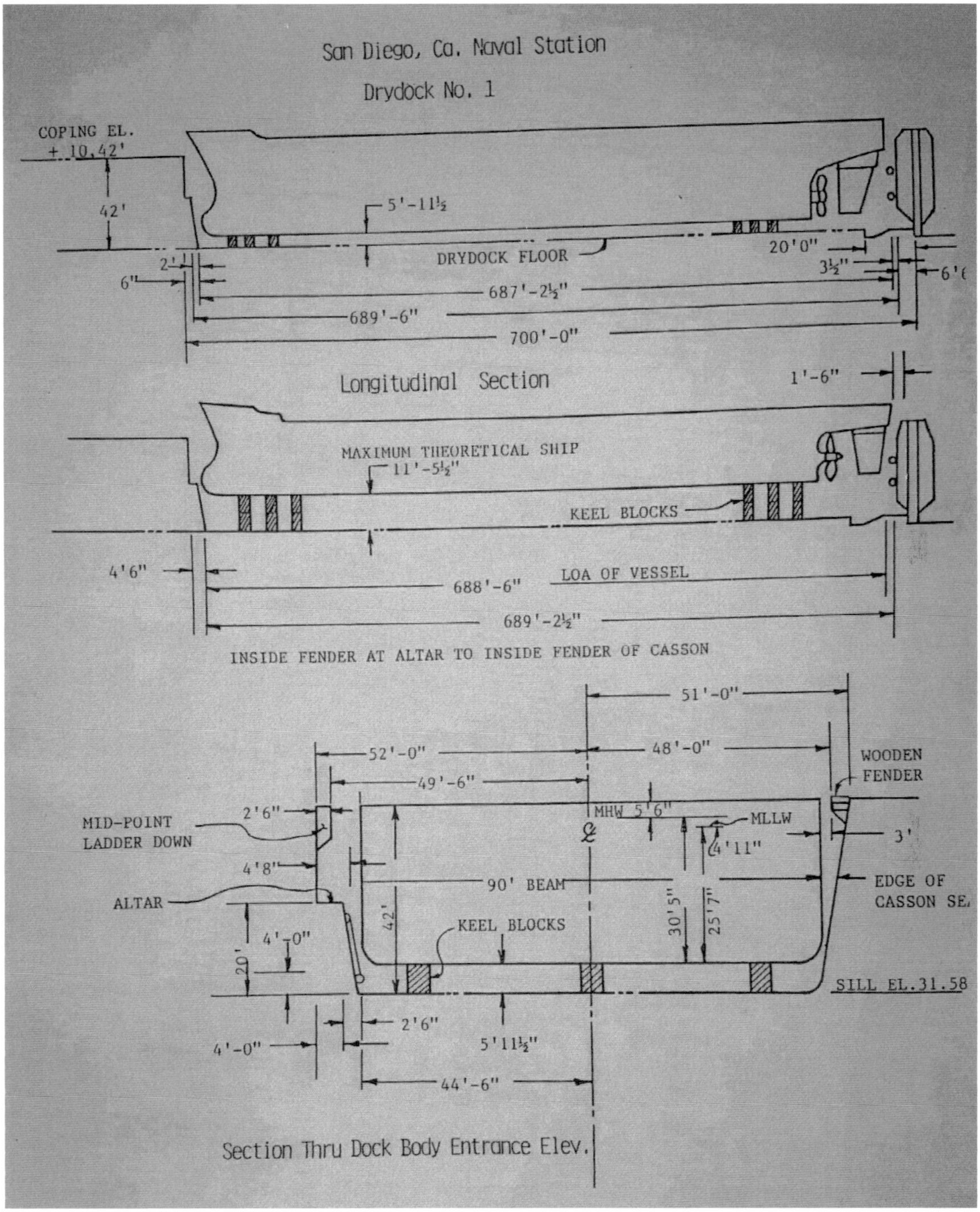

Figure A.3: San Diego Graving Dock Information Sheet page 2/2

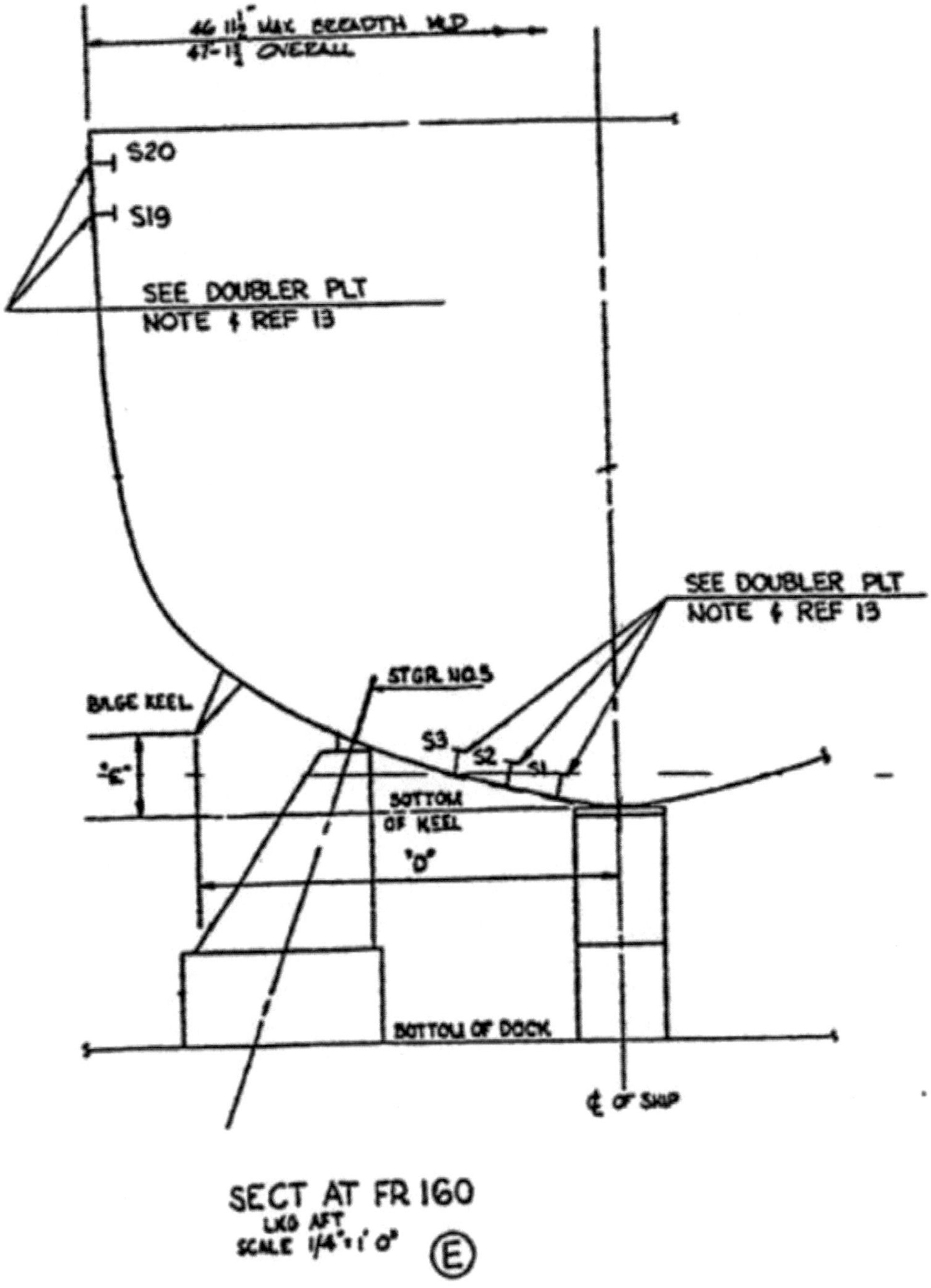

Figure A.4: Section AT FR 160 Diagram

Appendix B: Dimensioned CAD Components

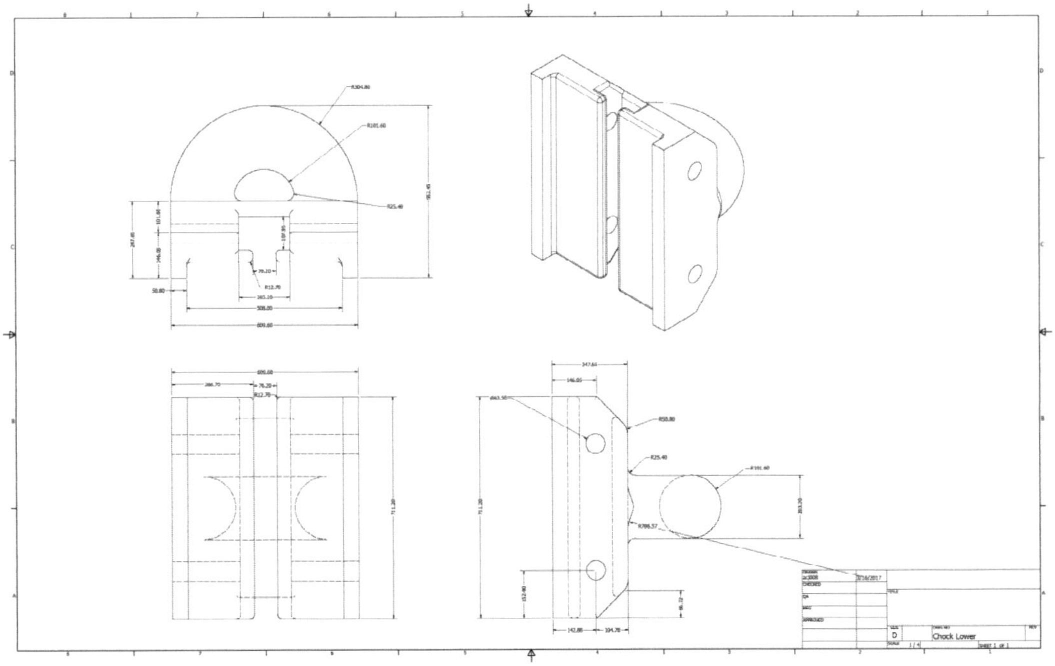

Figure C.1: Dimensioned CAD of Lower Chock

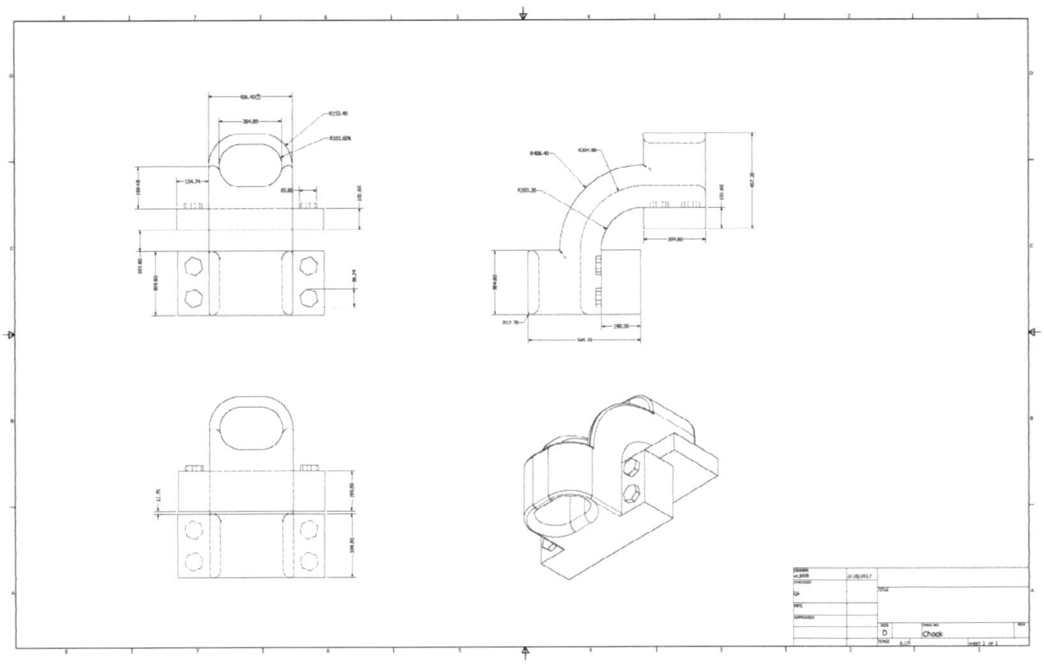

Figure C.2: Dimensioned CAD of Upper Chock

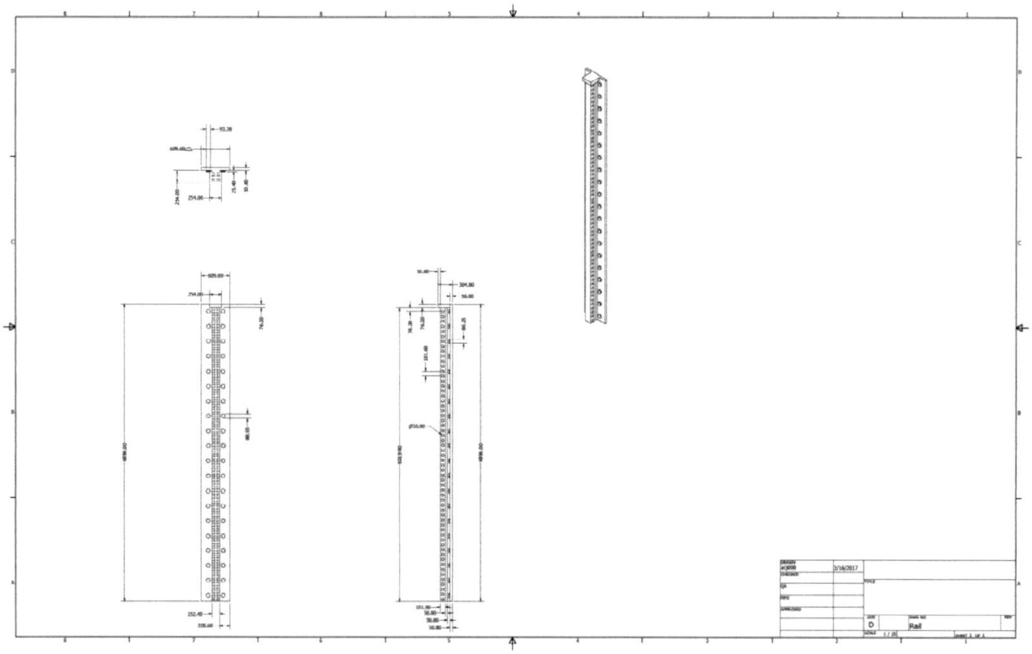

Figure C.3: Dimensioned CAD of Rail

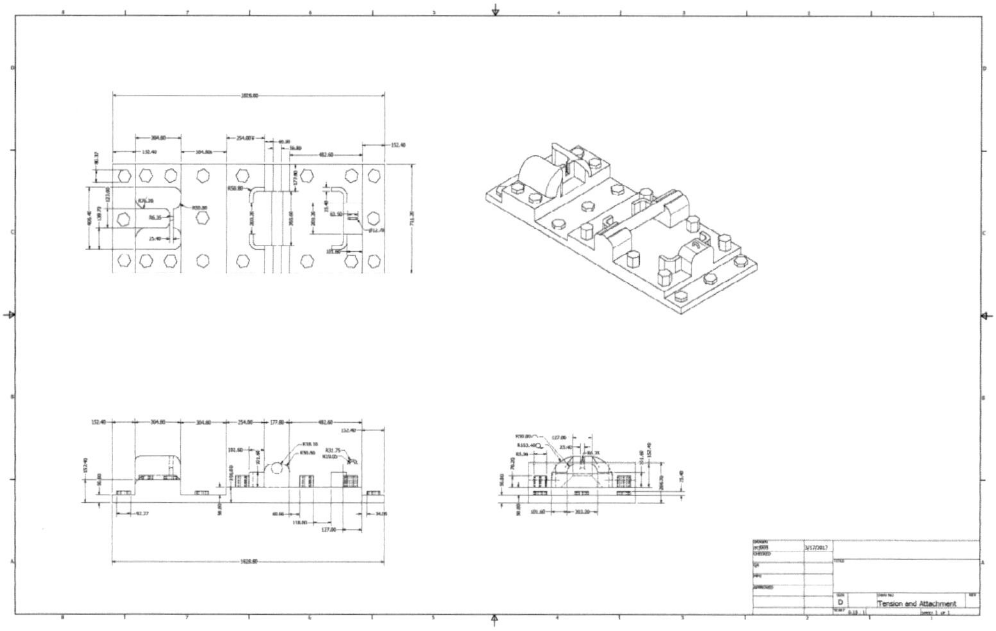

Figure C.4: Dimensioned CAD of Tension System

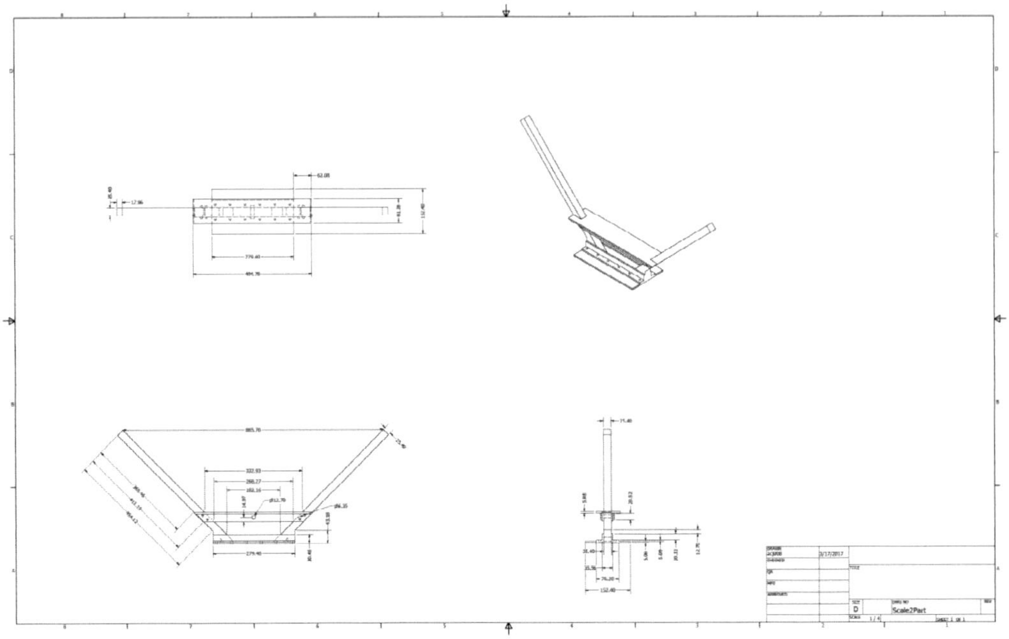

Figure C.5: Dimensioned CAD of Vibrations Test Platform

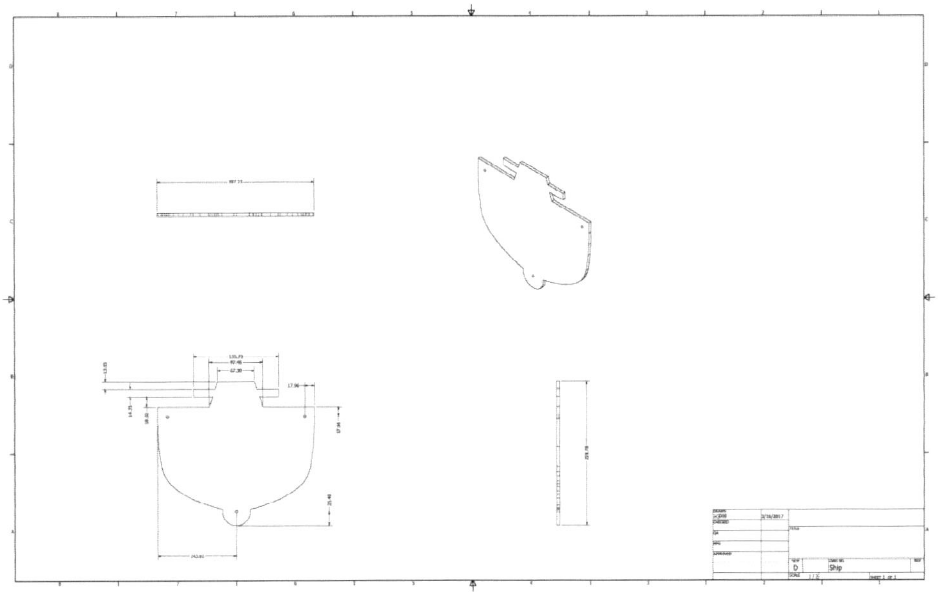

Figure C.6: Dimensioned CAD of Vibrations Test Platform Ship Cross Section

Appendix C: Full Cost Analysis

	Side Blocks		Chain Docking	
	Navy	Commercial	Navy	Commercial
Average dry docks per year:	7	16	10	24
	Side Blocks		Chain Docking	
	Skilled($/hour)	Unskilled ($/hour)	Skilled($/hour)	Unskilled ($/hour)
Wage	25	12	25	12
Overhead (100%)	50	24	50	24

Dry Docking with Side Blocks				
Startup materials	**Lifetime (yr.)**	**Cost (k$)/Lifetime**	**Navy**	**Commercial**
Softwood (reused)	1	3		
Wood replacement on blocks	10	1		
Hardwood	10	5		
Standard blocks(x150)	25	450		
Startup cost (k$):			459	459
Material/Equipment				
Softwood (cut to shape)	Single use	2.5		
Bracing	Single use	1		
Fasteners	Single use	2		
Crane	Single use	4		
Fork truck	Single use	2		
Tugs	Single use	3		
Miscellaneous	Single use	10		
Material/Equipment cost (k$)/year:			171.5	392
	Hours/dry dock			
Labor	**Skilled**	**Unskilled**		
Cutting blocks to shape	40	0	2	
Placement of keel blocks	40	80	3.92	
Placement of side blocks	40	80	3.92	
Tying down keel blocks	0	80	1.92	
Tying down side blocks	0	80	1.92	
Pilot	2	0	2	
Labor cost (k$)/year			109.76	250.88
Docking Masker Salary			80	80
Total Cost/Year			364.26	725.88
Total Cost/Docking			22.76625	103.6971429
Total Profit/Docking			6.829875	31.10914286
Total Profit/Year			109.278	217.764

Chain Docking

Startup materials	Lifetime (yr.)	Cost (k$)/Lifetime	Navy	Commercial
Excavation	Lifetime	1000		
Stud-link Chains (1' Diameter)(30ft)(x56)	25	19.6		
Rail system (x56)	25	112		
Shackle (x56)	25	0.56		
Close Panama Chock (Lower Chock) (x56)	25	161.28		
Upper Chock (x56)	25	53.76		
Hand winch	25	1		
Standard blocks(x75)	25	225		
Softwood (reused)	1	1.5		
Wood replacement on blocks	10	0.5		
Startup cost (k$)			1575.2	1575.2
Material/Equipment				
Fasteners	Single use	2		
Crane	Single use	4		
Fork truck	Single use	2		
Tugs	Single use	3		
Miscellaneous	Single use	10		
Material/Equipment cost (k$)/year:			210	504
	Hours/dry dock			
Labor	**Skilled**	**Unskilled**		
Placement of keel blocks	14	25	0.749	
Tying down keel blocks	0	25	0.049	
Pilot	2	0	2	
Labor cost (k$)/year:			27.98	67.152
Docking Masker Salary			80	80
Total Cost/Year			319.48	652.652
Total Cost/Docking			13.31166667	65.2652
Total Profit/Docking			13.44808333	58.01150286
Total Profit/Year			322.754	580.1150286

Appendix D: Full FMECA Analysis

Description		The purpose of every subcomponent in our system	The different forms in which every subcomponent can fail.
Components	**Subcomponents**	**Functionality**	**Failure Modes**
Ship side attachment	Bitts	Mechanism to which the chain is attached to	Bitt fails due to rust/corrosion
""	Attachment/Shackle	Chain locking device	Chain attachment/shackle failure/fracture
Chain	chain	Supports ship	Chain failure/rupture
Routing system	Rail	Changes the height of the lower chock	Corrosion/rust
""	""	""	Material fracture
""	Lower chock	Changes angle at which the chain approaches the ship	Corrosion/rust
""	""	""	Material fracture
""	Upper chock	Changes the direction of the chain from vertical to horizontal	Material fracture
Dock side attachment	Tension system	Supports the chain on the dock	Inability to lock the chain into position
""	Spring/damper system*	Dampens any forces applied to the ship and allows for minimal chain displacement	Damper fluid leak
""	""	""	Spring loses restoring force
""	Concrete anchor	Anchors the tension system and spring/damper system to the dock	Anchor fracture
""	""	""	Concrete failure

The impact that any given failure would have on the overall system. A lower number means that the result of failure would be more catastrophic.	The estimated possibility of any given failure occurring. The number of any given subcomponent used in our system was taken into account, since having a greater number provides more opportunity for failure.	This section takes into account the difficulty of detecting any given failure.	The complexity of fixing any given situation. Time and cost were also taken into account	The value of criticality of any given failure. Takes into account severity, occurrence, detection and difficulty fixing. The lower the value, the more critical it is.
Severity (50%)	**Occurrence (15%)**	**Detection (30%)**	**Difficulty Fixing (5%)**	**Criticality**
5	4	5	4	4.8
1	5	3	5	2.4
1	4	1	5	1.65
4	1	5	4	3.85
4	5	4	3	4.1
4	1	5	4	3.85
4	5	4	3	4.1
5	5	4	3	4.6
2	4	2	3	2.35
1	4	5	3	2.75
3	5	2	3	3
2	5	1	1	2.1
2	5	5	1	3.3

These are the corrective actions that must be taken immediately after a failure occurred. These actions are considered emergency cases because it's assumed that the dry docking process has begun and the failure is barely being detected.	These are the results of implementing the emergency corrective actions.	These are tasked that can be performed in order to prevent a failure.
Immediate corrective actions (Emergency case)	**Effects (Emergency case)**	**Mitigation**
Use alternate bitt or ship attachment	limits ship onboard anchoring locations	Routine painting and maintenance, weld inspection
Install new u bolt	lose 1 chain	Regular inspection of component
Replace chain	lose 1 chain	Regular inspection of component
Route directly through upper chock and re-tension	Change in transverse angle	Painting and maintenance, inspection, rinse after salt water exposure
Route directly through upper chock and re-tension	Change in transverse angle	Inspection
Route directly through upper chock and re-tension	Change in transverse angle	Rinse after salt water exposure
Route directly through upper chock and re-tension	Change in transverse angle	Inspection
Attempt to secure chain to another attachment	Change in longitudinal angle	Inspection
Attempt to secure chain to another attachment	Change in longitudinal angle	Inspection
Attempt to secure chain to another attachment	Change in longitudinal angle	Inspect for wetness around seals
Attempt to secure chain to another attachment	Change in longitudinal angle	Inspect component
Attempt to secure chain to another attachment	Change in longitudinal angle	Inspect component
Attempt to secure chain to another attachment	Change in longitudinal angle	Inspect for cracks in cement

Once the failure is detected, these are action that can be taken in order to reduce its severity. These actions are temporary solutions and further actions should be taken	These are more permanent corrective actions that would completely eliminate a failure mode.	These are redundancy options to minimize a failure's criticality.
Corrective actions	**Alternate corrective actions**	**Precautionary options**
Apply additional supports (wale shores)	Weld on new Bitts	Attach to secondary bitt
Replace the attachment	Replace the attachment	Can add an redundant attachment
Replace chain	N/A	Use chain sleeves to provide additional support/security
Treat the rust	Replace spring/damper system	N/A
Welding	Replace spring/damper system	Can add anchor chain
Treat the rust	Replace Chock	N/A
Welding	Replace Chock	Can add anchor chain
Welding	Replace Chock	N/A
Welding/clear obstructions	Replace Tension system	Add back up chain attachment
Stop leak and refill damper	Replace spring/damper system	Add back up locking in event of spring damper failure
Replace spring	Replace spring/damper system	Add back up locking in event of spring damper failure
Welding	Excavate and replace anchors	Add redundant anchoring
Fill in cracks	Excavate and refill	Add redundant anchoring

Summary most to least critical	
Component	Criticality
Chain	1.65
Dock side attachment	2.7
Ship side attachment	3.6
Routing system	4.1
*Spring/Damper system should only be used in three cases.	
Case 1: Chain jerking force is too great	
Case 2: The necessary force to keep the system in tension is	
Case 3: Chain damping force is not sufficient	

Appendix E: Code

(These codes are written for MatLab)

Essential Catenary Code:

```
%----input----
Thetarc=[[-
0.5:0.1:0.4],[0.41:0.01:0.49],[0.491:0.001:0.499],[0.4991:0.0001:0.5]];
%Thetarc=[-0.5:0.1:0.5]; % [Deg] range of theta (rotating away from wall)
%Thetarc=[0.49:0.001:0.5]; % [Deg] range of theta (rotating away from wall)
Thetarcr=Thetarc*pi()/180; % Convert to radians

%----parameters----
hb=1.8288; % [m] height of  blocks
hbs=7.0628; % [m] half breadth of ship
%hd=12.192; % [m] height of attachment on dock  (right side) (must be heigher
than ship)
hsi=9.4488; % [m] initial height of attachment on ship (keel to main deck)
(left side)
di =15.24-hbs; % [m] initial distance

hd=hsi+hb-0.5
l=8.279
%l=di+0.1; % [m] length of chain
lambda=127; % [kg/m] mass per unit length
g=9.81; % [m/s^2] gravity constant

dx=0.001; % [m] step size

r=(hsi^2+hbs^2)^0.5;
Thetar=Thetarcr+atan(hsi/hbs);
dr=r*cos(Thetar);
dr=hbs-dr+di; % [m] range of d
hsr=r*sin(Thetar);
hsr=hsr+hb;% [m] range of h

Tsr=zeros(1,length(dr));
TsRAngleSr=zeros(1,length(dr));
TsRAngleDr=zeros(1,length(dr));
TsAngleSr=zeros(1,length(dr));
TsAngleDr=zeros(1,length(dr));

for i=1:length(dr)
        d=dr(i);
        hs=hsr(i);
        a=abs(hd-hs); % [m] difference in height
        if a==0
                a=0.0001;
        end
        if sqrt(d^2+a^2)>=l
                fprintf('Warning: Length is geometrically too short for the
gap at iteration i=%d',i)
                fprintf('\nLength increased by %d [m] to meet gap
size\n',(sqrt(d^2+a^2)-l))
                l=sqrt(d^2+a^2)+0.00001;
        end
        if (l^2-a^2)*log((l+a)/(l-a))/(2*a)>d
```

```
                syms h;
                eq=((l^2-a^2 )*(a + 2*h) - 2*l*(h*(a + h)*(l^2-a^2
))^(1/2))*(atanh(a^2/(l*(a + 2*h) - 2*(h*(a + h)*(l^2-a^2 ))^(1/2))))/a^2 ==
d;
                h=vpasolve(eq,h,1);
                l1=-(h*l-sqrt(h*(a+h)*(l*l-a*a)))/a;
        else
                syms h;
                eq=((l^2-a^2 )*(a + 2*h) + 2*l*(h*(a + h)*(l^2-a^2
))^(1/2))*(atanh(a^2/(l*(a + 2*h) + 2*(h*(a + h)*(l^2-a^2 ))^(1/2))))/a^2 ==
d;
                h=vpasolve(eq,h,1);
                l1=-(h*l+sqrt(h*(a+h)*(l*l-a*a)))/a;
        end
        u = 2*h/(l1*l1-h*h);
        x1=asinh(u*l1)/u;
        x=[0:dx:d];
        if (hd) > (hb+hsi)
                k=cosh(u*(-(d-x1)))/u;
                y=cosh(u*(x-(d-x1)))/u+hd-k;
                Tsr(i)=double(lambda*g*(l1^2+h^2)/(2*h));
                TsRAngleSr(i)=sinh(u*x1);
                TsRAngleDr(i)=sinh(u*(d-x1));
        else
                k=cosh(u*(-x1))/u;
                y=cosh(u*(x-x1))/u+hd-k;
                Tsr(i)=double(lambda*g*(l1^2+h^2)/(2*h)+((lambda*g*((l-
l1)^2+(h+a)^2)/(2*(h+a)))-(lambda*g*(l1^2+h^2)/(2*h)))/2);
                TsRAngleSr(i)=sinh(u*(d-x1));
                TsRAngleDr(i)=sinh(u*x1);
        end
        TsAngleSr(i)=double(atan(TsRAngleSr(i))*180/3.1416);
        TsAngleDr(i)=double(atan(TsRAngleDr(i))*180/3.1416);
        hold on
        plot(x,y)
        Tr(i)=double(lambda*g/u); %Tension in x-direction
        labels{i}=sprintf('T(Theta=%4.2f [deg]) = %0.f [N]', Thetarc(i),
Tsr(i));
end
legend(labels)
xlabel('Distance [m]')
ylabel('Height [m]')
title(sprintf('Graph of Chain Geometry vs Varying Ship Angle Given Fixed
Length of L = %0.1f [m]', l))

figure(2)
hold on
plot(Thetarc,Tsr);
xlabel('Change in Ship Rotation [deg]')
ylabel('Tension [N]')
title(sprintf('Graph of  Tension vs Varying Theta Given Fixed Length of L =
%0.1f [m]', l))

Results(:,1)=Thetarc.';
Results(:,2)=Tsr.';
Results(:,3)=TsAngleDr.';
Results(:,4)=TsAngleSr.';
```

```
fprintf('%0.2f %0.f %0.1f %0.1f\n', Results');
```

Vibrations Code

```
https://www.mathworks.com/help/matlab/ref/ode45.html?requestedDomain=www.math
works.com

[t,p] = ode45(@SpringTwo, [0 20], [0; 0]);
plot(t,p(:,1),t,p(:,2))

-----------------
function [ pdot ] = SpringTwo( t,p )

m = 3566321.741; % [kg]
%Is=2042.408*(0.3048^4)*(2*m/79.515); %Inertia Tensor x mass/volume, mass
moment of inertia [m^4]
cg=5.7912; % [ft] Distance Keel to Center of gravity
Is=0.5*m*7.0628*7.0628+m*cg*cg;
g=9.806;
rad=11.7967;
d0=8.1772;
d=d0+rad*p(2)
l=8.3;
lo=l;
a=0.00000001;
rhoAg=127; % [kg/m] mass per unit length

T=0.2*m;

syms h;
if sqrt(d^2+a^2)>=lo
                l=sqrt(d^2+a^2)+0.0001
end
if (abs(a) > 1.e-6)
    b=sqrt(h*(a+h)*(l*l-a*a));
    c1=a*a/(l*(a+2*h)-2*b);
    c=atanh(c1);
    e=((l*l-a*a)*(a+2*h)-2*l*b)/a/a;
    y=e*c-d;
    ht=vpasolve(y,h,1);
    htt=double(ht);
    l1=-(htt*l-sqrt(htt*(a+htt)*(l*l-a*a)))/a;
    mu = 2*htt/(l1*l1-htt*htt);

else
    b=log((l/2+h)/(l/2-h));
    y=(l*l/4-h*h)*b/2/h-d/2;
    ht=vpasolve(y,h,1);
    htt=double(ht);
    mu = 2*htt/(l*l/4-htt*htt);
end

f1=rhoAg/mu;
%

d=d0-rad*p(2);
if (abs(a) > 1.e-6)
    b=sqrt(h*(a+h)*(l*l-a*a));
```

```
    c1=a*a/(l*(a+2*h)-2*b);
    c=atanh(c1);
    e=((l*l-a*a)*(a+2*h)-2*l*b)/a/a;
    y=e*c-d;
    ht=vpasolve(y,h,1);
    htt=double(ht);
    l1=-(htt*l-sqrt(htt*(a+htt)*(l*l-a*a)))/a;
    mu = 2*htt/(l1*l1-htt*htt);
else
    b=log((l/2+h)/(l/2-h));
    y=(l*l/4-h*h)*b/2/h-d/2;
    ht=vpasolve(y,h,1);
    htt=double(ht);
    mu = 2*htt/(l*l/4-htt*htt);
end

f2=rhoAg/mu ;

extF=(l-lo)*3.14*2*14*9*0.0254*0.0254*1000000000/g;

kr=(extF+f1-f2)*rad;

c=2000;
c=0;

%pdot=[0;0];
pdot=[(m*cg*g*sin(p(2))-kr+T+-c*p(1))/Is; p(1)];

%pdot= [(m*cg*g*sin(p(2))-kr+T-c*p(1))/Is ; p(1)];
%pdot= [p(2) ; (m*cg*g*sin(p(2))-kr+T)/Is];


end
```

Why Do Dry Docks Sink and Not Come Up

Robert (Bob) G. Wasalaski
4-5 May 2017
International Dry Dock Conference
Houston, TX

Prologue

- To know the future, study the past.
  - Those who don't are subject to repeat it.
    - A student retains about 80% of what is taught.
  - Man's progress can be tracked to the ability to measure
    - As measurement became more process
    - It became repeatable
    - Repeatability lead to more accuracy.
- Docking accidents/incidents which have set criteria
  - Some date to the beginning
  - Some date to the 1930s and 1940s
  - Some are new
  - Some have been relearned

9/12/2017

2

Background

- How did I get into this type of work?
  - Naval Architect and Marine Engineer by education
  - Naval engineer by training
    - Docking Engineer
    - Docking Officer
  - By career
    - Experience is the best teacher
      - Ingalls Launch Pontoon accident
      - Ballasting of US Navy dry docks
    - NAVSEA Docking Technical Expert
      - Support unusual docking
      - Investigate docking incidents involving USN ships
    - Internationally recognized expert for
      - Docking
      - Heavy lift transport
      - Stability of ships

Incidents can be Devastating

Ingalls Launch Pontoon accident
during undocking of DD-963

- Scenario
  - Dock started to bend in middle (sag)
  - Center of pontoon deck started to flood
  - Cracking sounds heard
  - Two moveable wingwall sections fell ove
  - Dock listed and trimmed as tanks floode
  - Strength compromised
    - Unable to re-lift ship on one half dock
    - Dock ballasted down to let ship off

- Ship slid aft in dock
- Holes punched in pontoon deck
- Safety deck not subdivided
- Loss power
- Dock sank

Post Accident Evaluation of Ingalls Launch Pontoon

- Pontoon broke in half with DD-963 on board
  - Placement of ship on dock over critical midship joint
  - Pumping plan not realistic
    - Based on weight curve of ship
  - Monitoring of Ballasting of dock to ship load was confusing
    - flashing digital lights for 48 tanks
  - Dock bending caused shear on wing wall sections causing
    - Seals to leak
  - Modifications to dock
    - Allowed wing wall sections to move
- Controlled sinking of Dock out from under ship
  - Ship slides aft
    - Damages (holes) pontoon deck causing additional flooding
  - Lack of watertight bulkheads on safety deck in wing wall causes
    - loss of power
- Salvaged, repaired and extensively modified

Affects on Criteria

- Placement of ship
  - Pumping plan to account for loading on dock
- Seal failure impacts seepage rates
  - Flooding in excess of pump capacities
- Structural integrity of dock
  - Longitudinal strength-bending
    - Include monitoring of temperature and density
  - Differential heads
  - Monitoring of dock deflection during operations
- Watertight bulkheads on Safety deck
  - Segregation of power generators and pumps
- Limit of trim after damage to avoid ship from sliding on blocks
  - 3 degree or 20 feet trim after damage limit
- Height of air vents
  - Set depth of submergence
  - Freeboard requirement on wing wall

MEASURE, MARK, CHECK

- The pumping plan is your guide.
- Tank level indicators are you tool.
- Deflection monitoring is your visual check.
  - Impact of temperature and density

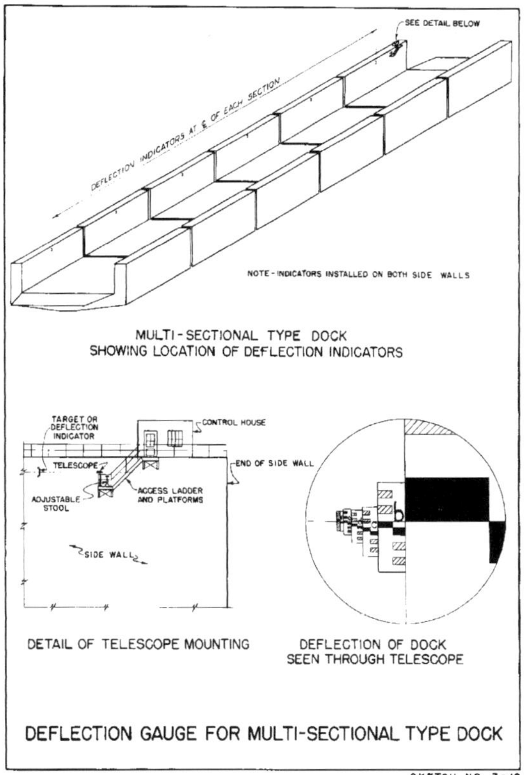

SKETCH NO. 7-18

Investigation called for by

OSHA® Occupational Safety and Health Administration

www.osha.gov

Shipyard Industry Standards

Accident and Incident Investigation

A careful and thorough analysis will identify conditions and actions that contributed to or led up to the incident. Once the causal factors are revealed, the employer can make the necessary corrections to prevent recurrence.

Is incident caused by A human error, A Procedural error or A design error.!

Marine Forensic Investigation

- Marine- of ships and vessels
- Forensic- of accidents and wrecks
- Investigation- to determine causes

Investigation

- OSHA called for investigations are to establish blame
  - Historical trend of accidents/incidents
    - 80-90% of accidents are caused by Human error
      - Mis-judgement
      - Physical problems with the operation
    - 10-20% of accidents are caused by
      - Deficiencies of the facility
      - Deficiencies of documentation and/or training
- MFI to determine Deficiencies and Procedural Shortcomings
  - Improve the criteria
  - Improve the facility
  - Improve the operation

To Begin-Floating Bodies

- **Archimedes** 287-212 BC
  - One of the greatest scientists and mathematicians of the ancient world and all time.
    - His studies of shapes and relationships revolutionized Geometry.
    - Recognized as the father of mathematical physics and Integral Calculus.
  - The story,
    - A crown had been made for King Hiero. Archimedes was asked to determine whether the goldsmith had made the crown of pure gold.
    - While taking a bath, he noticed that the level of the water in the tub rose as he got in.
    - Realizing that this effect could be used to determine the makeup of the crown, Archimedes jumped out of the tub and rain into the street yelling "Eureka!" (Greek: "ε ρηκα!," meaning "I have found it!").
    - So excited by his discovery that he had forgotten his clothes.
  - Father of theoretical naval architecture.
    - Three discoveries that affect the workings of floating bodies/vessels
    - These effects , known as Archimedes' Principle, described in his treatise *On Floating Bodies*.

9/12/2017 12

Archimedes - Rule of Buoyancy

- A body floats at a submerged volume equal to the weight of water displaced
- Placing the crown and a block of gold on identical boxes of wood,
- If the block of gold and the crown are equal, the two blocks of wood sink to the same level.

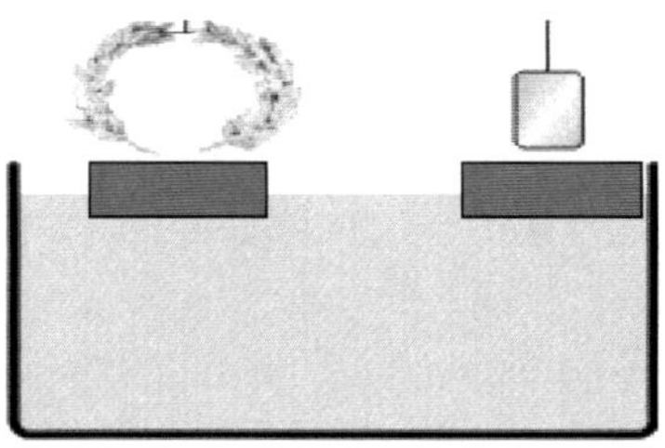

- He called this uplifting force *Buoyancy*.
- A ship on a floating dry dock is similar to the crown on the block of wood.

9/12/2017 13

Buoyancy=Displaced Volume

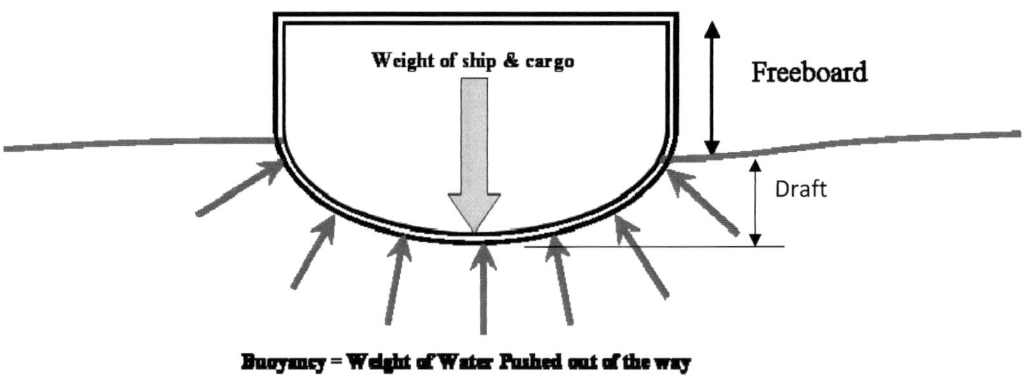

So the weight of the ship pushes the water out of the way.

Weight=Displaced Volume of the ship

Displacement=length x beam x draft x form coef/ density of water

9/12/2017 14

History of Dry Docks

- The dry dock was invented in Ptolemaic Egypt,
  - A Phoenician devised a new method of launching it (a ship), having dug a deep trench under it, equal to the ship itself in length, which he dug close to the harbor. And in the trench he built props of solid stone five cubits deep, and across them he laid beams crosswise, running the whole width of the trench, at four cubits' distance from one another; and then making a channel from the sea he filled all the space which he had excavated with water, out of which he easily brought the ship by the aid of whatever men happened to be at hand; then closing the entrance which had been originally made, he drained the water off again by means of engines (organois); and when this had been done the vessel rested securely on the before-mentioned cross-beams.[1]
- The first modern European[4] and oldest surviving dry dock still in use at HMNB Portsmouth.
  - Commissioned by Henry VII of England in 1495 (see Tudor navy).
  - Dry dock currently holds the world's oldest commissioned warship, HMS Victory.
- First US dry dock is gravy dock #1 at Norfolk Naval Shipyard
  - Christened by John Quincy Adams in 1803.
- First floating dry docks ordered for the US Navy was in 1847

Graving Dry Dock

- A trench in the shore
- With a capability to close the end
- Ship displaces water
- Remaining water is pumped out
- Main Issues
  - Blocking
  - Ship stability
  - Clearance
    - Tide and water density
  - Flooding

Floating Dry Dock

- Floating dry dock- a U shaped vessel
  - Divided into tanks
    - Flood valves allows water in
    - Pumps pushes water out
  - Weight of the ship displaces the water in the dock

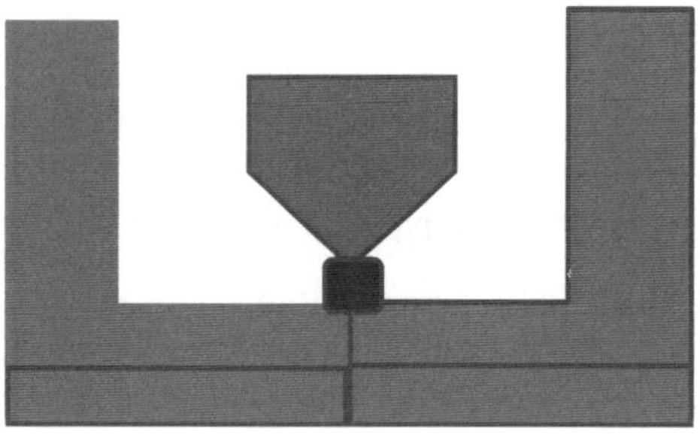

- Water pumped out
  - Equates weight of ship
  - Center is under center of gravity of ship

Basically, the repeat of Archimedes' first principle!

Early Floating Dry Dock

- Description of floating dry dock in Venice in 1560
  - Use: new method to salvage a grounded ship

Docking Incidents

- Weight of ship is greater than buoyancy under it
  - Over stress structure
  - Differential head between tanks over stress bulkheads
- Ship weighs more than lift capacity (buoyancy)
  - Dock will not be able to lift the ship out of the water
  - Loss of freeboard-Reserve Buoyance

Ship's Weigh Curve to Dock Loading

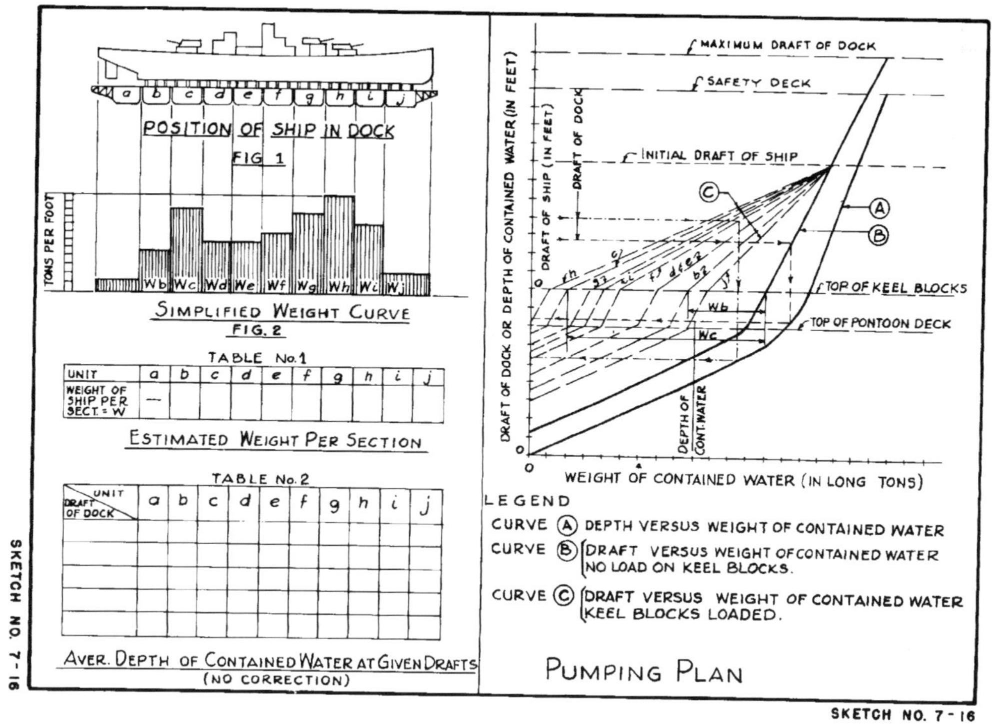

ARDM-4

- Tank arrangement and wing wall openings cause dock to bend and twist during ballasting
- Movement of cranes from amidship position causes dock to trim to end where cranes have been re-located
  - 1.5 ft of trim in the deballasted condition
  - 3 ft of trim in the ballasted down condition
- Affects
  - Strength
  - Stability
  - Trim
  - Complicates operation

Unburdened Dock Pumping Plan

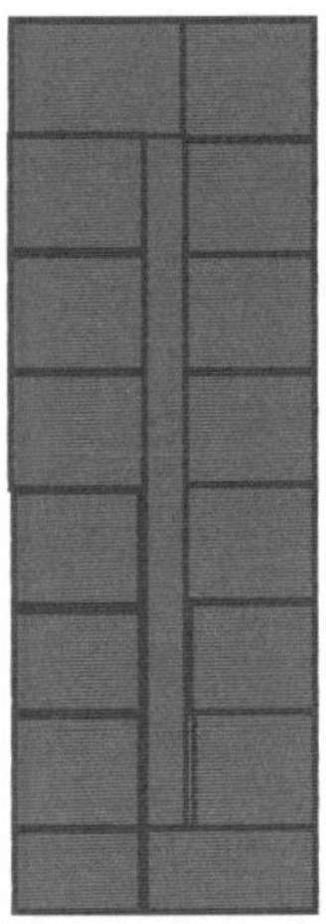

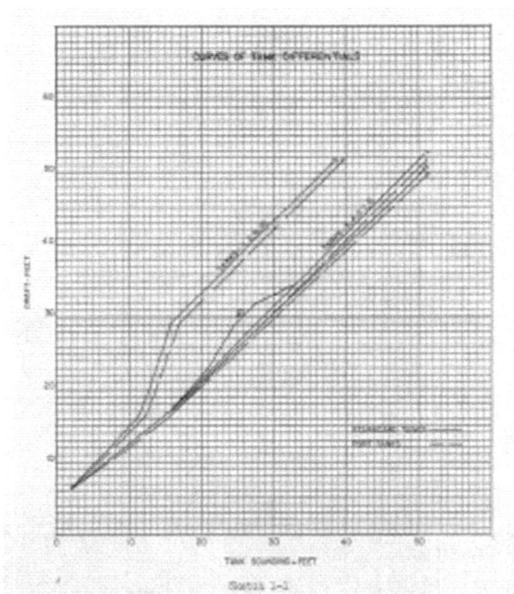

Calibration of a floating dry dock involves matching the tank capacity curves to the afloat drafts of the dry dock.

Affects on Criteria

- Criteria for
  - Ballasting (pumping) plans
  - Calibration of dock
    - Tank/Draft capacity curves
  - Operational limitations on movement of cranes

Cruiser Drydocking in AFDM

- Ships
  - CGN emergency dry docking
  - CG-47 class
- Short high weight ships
  - Heavily load one end of the dock
  - Short ships stress docks

What is the capacity of the dock?

MIL-STD-1625B(SH)
31 August 1984

FIGURE 9. Lifting capacity variation with ship location.

- Buoyancy to accommodate the ship's weight,
- Longitudinal strength to accommodate the deck loading,
- As center of gravity of ship moves off center of dock/blocking
    - Loading to one end increases
    - Lifting capacity decreases
        - Buoyancy average deceases
        - Structural loading average decreases

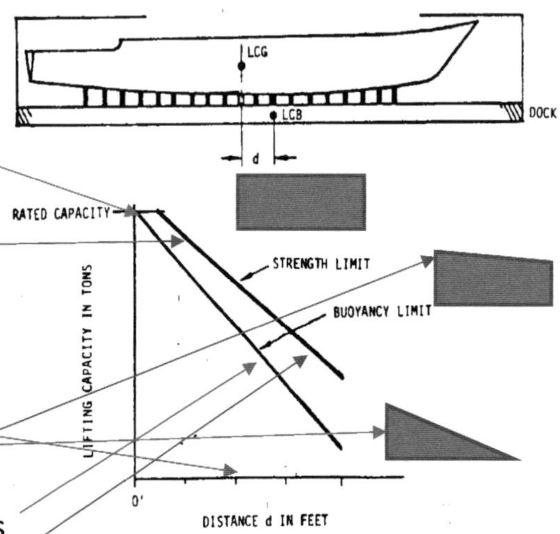

Differential Loading

MIL-STD-1625B(SH)
APPENDIX D
31 August 1984

- Trapezoidal Method as shown in MIL-STD-1625

    If A<B
    Lmax=W*(1+A/B)/Lk
    Lmin=W*(1-A/B)/Lk

- Used to assess:
    Block loading
    Deck loading
    Tank loading

- Length of overhang may give higher loads than predicted.

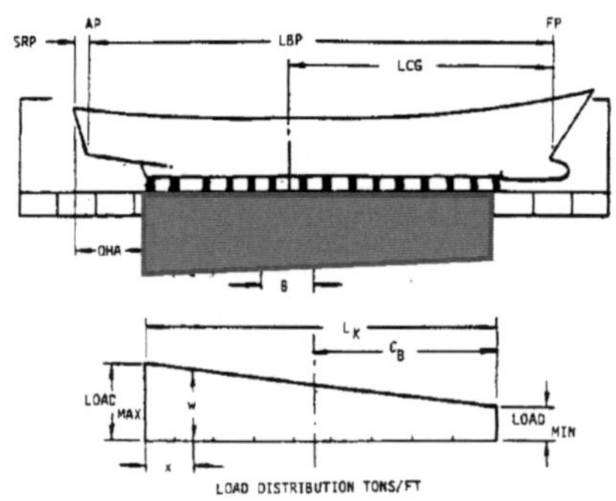

LBP = length between perpendiculars of ship

SRP = distance from after perpendicular (AP) to point from which distance to keel blocks is measured

LCG = distance from forward perpendicular (FP) to ship's longitudinal center of gravity

OHA = length of overhang from SRP to first after keel block

W = displacement of ship

L_K = length of keel blocking

$C_B = \frac{L_K}{2}$ = distance from the end of keel blocking to the center of blocking

$B = \frac{L_K}{6}$ = distance from center of blocking to the approximate center of the loading trapezoid

$A = C_B$ - (LBP+SRP-LCG-OHA) = distance from ship's LCG to the center of blocking

c = length of keel blocking over a tank

w = average loading over a tank

x = distance from the after end of keel blocking to the center of gravity of a tank

FIGURE 18. Ship load distribution on dock.

98

Block Loading Analysis Prepared For Ships With Long Overhangs

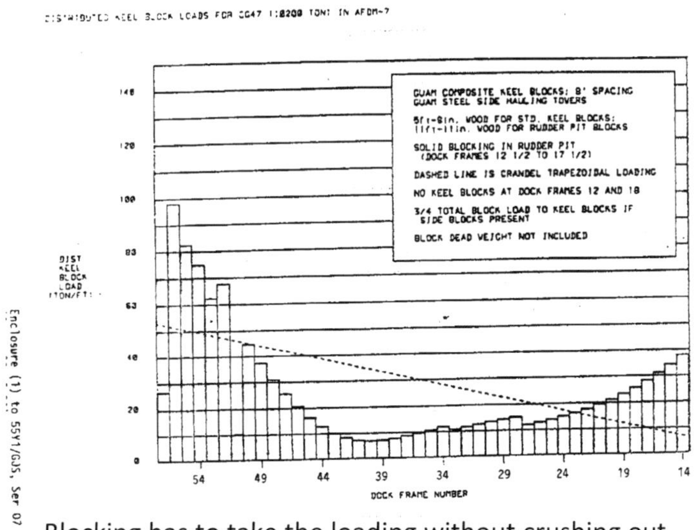

Blocking has to take the loading without crushing out
Blocking disperses the load to the dock deck structure
Dock structure has to take both the local and longitudinal loading
Sometimes actual loading is greater than estimated
Available buoyancy has to account for all of this.

Loading on Blocking and Dock Deck

- Soft wood caps disperse load on blocks.
- Blocking disperse the load to the deck
  - Spreader beams used to spread the load on dock floor

Note steel plate on top of soft wood caps to spread load on blocks.

Local Deck Loading

- If dock floor can not support load, spreader beams can be used to disperse, spread the load.
  - Under keel blocks
  - Under side blocks

Note spreader beams in transverse and longitudinal directions.

Affects on Criteria

- Criteria for
  - Capacity limit change
    - Tons per foot of loading limit
    - Lifting capacity curves

CGN dry docked in AFDM

- AFDM had a large sonar pit at one end
  - Freeboard of pit to avoid flooding
  - Trim after damage to dock
- Trimmed dock to avoid flooding pit

Freeboard

Freeboard for
- Density change
- List or trim allowance
- Wave action
- Reserve buoyancy

Docking Incidents due to Structural Failure

- Ship's weight is more than lift capacity
  - Loss of freeboard
  - Loss of reserve buoyancy
  - Differential pumping may overload bulkheads in dock or shell of the dry dock
- Ship's weight is more than the structural capacity
  - Local overloading stresses/punches holes in the deck
  - Overall loading causes the dock to bend or break

Example: AFDL-47

- Heavy concentrated load over center tanks
  - Not adequately compensated for by ballast water removal in end tanks
    - Differential head stress on bulkheads
      - Between tanks
      - Hull shell
      - Cracking of wing wall along bulkheads and vents
    - Longitudinal bending of dock
      - Dock structure takes up load
      - Deflection in the dock
- Pumping unloaded end tanks causes
  - Further stress and bending in dock
  - Wrinkling of side shell, pontoon deck, wing wall decks

Affects on Criteria

- Criteria change
  - Pumping plans
  - Freeboard
- AFDL 47
  - Taken out of service
  - In second hand service at reduced Capacity

Norfolk Dry Dock Incidents-3

- In 2000 a 14000 ton floating dry dock experienced substantial structural failure
  - Overload of center ballast tanks
    - Insufficient compensated by ballast water removal
    - Causing transverse bending of dock
  - Pumping unloaded wing tanks causes further bending of dock
    - Causing further stress and bending in dock
    - Causing bulking of pontoon deck
  - Contributing factors
    - Deck corrosion contributed to weakening of the pontoon deck structure
    - Ripple affect
    - As the deck started to fail under the highest loading, other sections became overloaded and cause further failure.
- Similar accident happened on two other floating dry dock

Large ships overstress the dock structure.
Corrosion may lessen capacity.

Don't forget transverse loading!

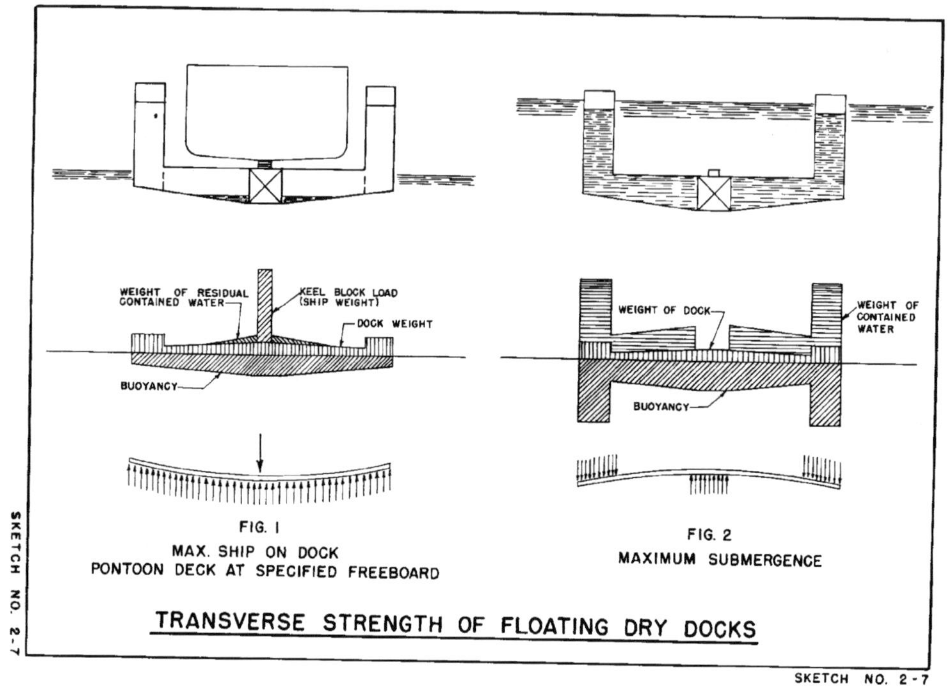

Transverse structural also has a load and buoyancy limit.
Drawing from 1940s Docking Manual!

Affects on Criteria

- Criteria affected
  - Pumping plans
    - Consider transverse loading
  - Structural requirements for transverse loading
  - Hull material inspection of pontoon decks

Archimedes 2nd Principle- Rule of Stability

- Placing the crown and the block of gold suspended from a scale on identical boxes of wood,

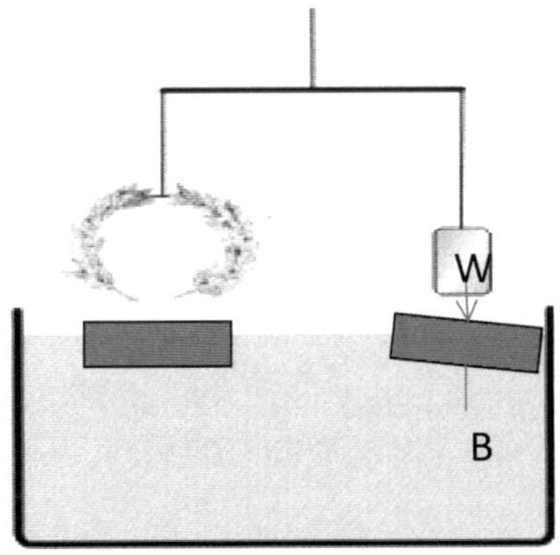

•If the crown and block of gold are of the same weight but of different centers, the scale will also tip to that side.
•The block of would will tip until the center of buoyancy is under the center of the line of action of the weight acting down
•This shifting of uplifting force *Stability*.

9/12/2017
39

Stability changes with draft of the dock

- The stability of the dock changes as
  - It picks up the weight of the ship
  - The ballast water is pumped out of the dock
  - The shape of the dock changes
- Stability is measured by GM
  - Maximum when pontoon deck out of water
  - Minimum with the water level at the keel block height
  - Change in GM affects the amount of list and trim-Multiplication factor

FIGURE 4 VARIATION OF STABILITY DURING DOCKING EVOLUTIONS

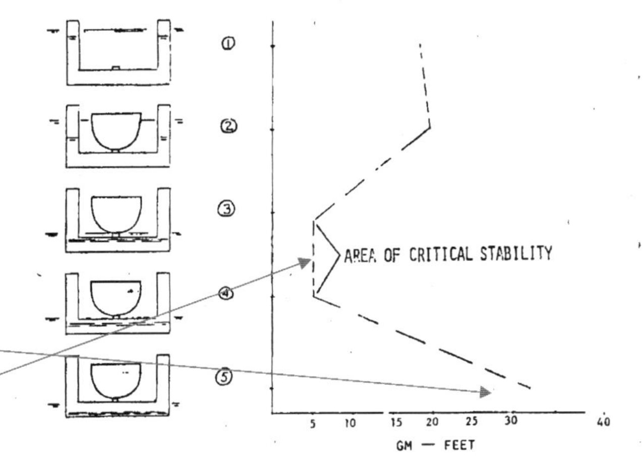

Multiplication Factor

Multiplication Factor, fm:

- I.E. GM changes vs list

• Fm=GM ratio	30/5	30/5	60/5	30/3	30/3
• list multiplied by	6	6	12	10	10
• list during ballasting down	1/4	1/2	1/4	¼	1/2
• list pontoon deck submerged	1.5	3	3	2.5	5

- Affected by
  - Change is shape of dry dock as draft changes and pontoon deck submerges
  - Change with relationship to separation of the wing walls
  - Change with shape and width of wing walls
  - Tank subdivision configuration and free surface effect in tanks
  - Freeboard to pontoon deck

Phase of minimum stability

Freeboard

Class requires

- Rated
- Minimum

USN Mil-Std-1625

- dependent on lift capacity

Often in commercial

- operations the docks operate with zero freeboard or less

Class and Mil-Std

- requires a Minimum GM value

Zero freeboard still common in commercial industry

YFD Failure to Lift
High Center of Gravity Ship

- Ship within capacity of dock
  - Dock starts to deballast
  - Dock lists and trims as it approaches point of minimum stability
  - Operation stopped safely
  - Other similar cases have resulted in
    - Dropped ships
    - Sunk or damaged docks
- Dock modified to increase stability in range of minimum stability

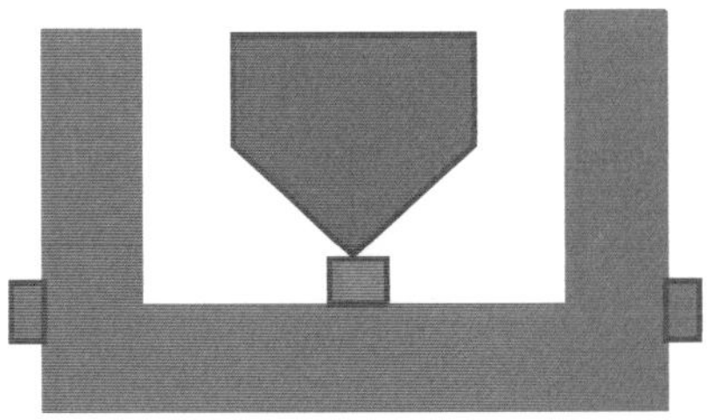

Floating Drydock Windwall Mods to Improve Stability in Minimum Plase

Some dry docks are designed to compensate for this issue.
Others are modified to compensate for the issue.

Affects on criteria

- Freeboard requirements set for both
  - Classification society
  - USN Mil-Std-1625
- Stability criteria changes
  - Stability criteria takes free surface into account
  - Stability criteria neglects affect of ship to stabilize dock
  - Change in criteria of when to position side blocks

Cruise Ship Docking Incident

- Possible causes
  - Shifting weight on a ship
  - Water seepage causing pontoon deck to submerged uncontrolled
- Dock began to list
  - Submerging of edge of pontoon deck
    - Causing change in stability
      - Change in GM as pontoon deck submerges
      - Free surface in tanks further decreases GM and stability
    - Causing change in list and trim
- Overload of side blocks
  - Blocks punch holes in deck
  - Causing further flooding of tanks in dock
  - Causing ship to fall against and crushing wing wall
- Dock sinks

Affects on Criteria

- Monitoring of weight shifts while ship in dock
- Assessing of loads on
  - Keel Blocks
  - Side blocks
    - Portion of ship load
    - List of ship/dock
    - Wind load
    - Seismic load
  - And dock deck

Archimedes' Screw

- Also called **screw pump**
  - A machine historically used for transferring water
  - Screw pump is commonly attributed to Archimedes
  - Used to remove water from the bilges of galleys
    - Also used to move water into irrigation ditches

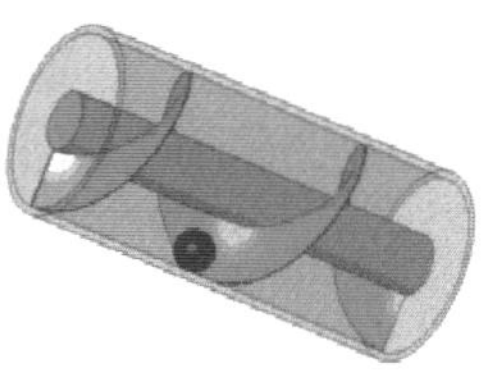

  - Basis of modern day pumps
  - Basis for modern ship propellers
  - How water is pumped from floating dry dock.

Affects on Criteria

- Criteria affected
  - Design, construction and repair
  - Hull material inspection of dock
    - Systems
      - Pumps and Piping
        - » Ballasting/deballasting times
      - Electrical power
  - Operating procedures
    - Watch rotation
    - Sound and security

Factors for Seepage and Sinkage

- Seepage/sinkage rates affects
  - Loss of control of flooding
    - Will flooding cause dock to list or trim
    - Is flooding greater than pumping rate
      - Can pumping control seepage/flooding/sinking
    - Freeboard
      - How long before pontoon deck submerges
  - Stability of dock
    - Is stability adequate to keep pontoon deck from submerging
      - What is the change in stability or list as pontoon deck edge submerge
    - Will vessel fall off blocks
    - Is stability adequate to keep dock from capsizing
  - Structure of dock
    - Structure of deck adequate to take extra load on side blocks
    - Flooding will not overstress the dock
      - Cause bending to exceed longitudinal strength
      - Cause local loading on shell and bulkheads which overstress the structure
  - Watch spacing
    - Is freeboard adequate for the watch rotations
      - Sound and security check often enough to observe and control flooding in time
    - Is pumping rate adequate for watch rotation
      - Can pumps control flooding

Affects on Criteria

- Hull material inspection to assess wastage on
  - Local strength
  - Overall strength
  - Differential heads of
    - Sides of docks
    - Bulkheads of dock
  - Underwater hull inspection
    - Dry docking
    - Underwater inspection by divers
- HM&E inspection to assess maintenance and operation
  - Flooding and pumping systems
  - Flooding piping and valves
    - Air vents
  - Pumping system
  - Electrical power system
- Backup procedures
  - Manning
  - Watch shifts
  - Sound and security

Flood Valves, Vents and Pump

- Piping sizes
  - Adequate for operation
- Valves Operation
  - Easy to operate
    - Personnel friendly easy to understand operation
  - Opening/closing rates
    - Affects control of flooding
  - Can be clogged so won't close
    - Causing flooding
    - Causing over pressurization of bulkheads
- Pump systems
  - Can pumps match piping flood rate
  - Can Pumps back flood
    - Isolation or check valves not closed or clogged
- Vents
  - Set at right height for submerging depth
- What is an acceptable flooding/pumping rate?
  - If closing rate of valves is 20 minutes, is it okay
    - If dock submerges in 1.5 hours probably not
    - If dock submerges in 8 hours may be okay
  - If the flooding rate is controlled by pumping, is it acceptable
    - Does the flooding overstress the structure of the dry dock
  - <u>Can you control the operation of the dry dock!</u>

Factors to consider for Flooding Rates

- Flooding/sinkage rate affects
  - Pumping rate
  - Flooding control
  - Freeboard
  - Trim of dock
  - Stability of dock
  - Watch/shift spacing

Dock Flooding Incidents

- Seepage Rate Accidents
  - Applied mainly to Wooden dry dock
    - Need to ballast down to maintain hull wet to keep seals tight
    - Deballasted for long periods
      - dries out seams and seals
      - Causing seepage when submerged until seams tighten
      - Excessive seepages causes dock to flood
  - Applies to badly corroded steel docks
  - Applies to docks with old/poorly maintained
    - Seals for tank openings
    - Pumps and valves
- What is an acceptable seepage rate?
  - Example
    - If the seepage rate is 120000 gals an hour and the pumping rate is 150000 gals an hour, is it acceptable?
  - Must maintain control of the dock
    - Flooding so that dock does not sink
    - Pumping to exceed flooding
    - Stability so dock does not capsize

Dry Dock and Tug Capsize at Everett Shipyard

- March 18, 2012
  - On Saturday a (AFDL) dry dock containing the 140-foot tug Invader began sinking
    - Dock is at the Vigor shipyard on the Everett, Washington waterfront.
    - Tug fell over
    - Dock sinks by capsize
  - The Coast Guard and Dept. of Ecology have responded investigating the incident

Dry Dock /Tug Accident

photo courtesy U.S. Coast Guard

August 2013 Out of Service

Shipyard personnel blame faulty check valve for sinking.

Because it was the weekend, no one was aboard either Invader or the dry dock at the time of the accident.

Ship fell over in dock, block failure

Note patches on side of dock observed in second photo

ARDM-2

- Cranes not secured in forward position during ballasting down operation of the dock
  - Cranes roll to aft end of dock
  - Cranes hit stops and tip over the end of dock wing walls
  - Clips on crane rails pull rails off the deck opening the submerged wing walls
- Dock sank
  - Trim due to movement of cranes
  - Submergence of wing wall
  - Free surface of liquids in tanks

Affects on Criteria

- Criteria for
  - Stowage of cranes
    - Ballasting for where cranes are located
    - Do not move cranes during ballasting
  - Watertight integrity of safety deck
    - Compartmentation
    - Operating procedures

AFDM-2

- Hurricane winds broke dry dock loose from moorings
  - Collisions with barges and flood debris hole dock
  - As dry dock sank, it trimmed and listed
  - Ship in dock fell over
  - Dock lists further and capsizes
  - Ship floats off
  - Capsized dock grounds on river shoal
- Salvaged by US Navy
- Back in service in commercial industry in Texas today

Affects on Criteria

- Wind criteria for dry dock drove
  - Storm mooring
    - 100 knot beam winds
    - And current
  - Storm ballasting
  - Freeboard requirement
  - Stability of dock
    - 100 knots beam wind
  - Side block design
    - 100 knot beam wind

9/11/2001 Security Requirements

- Security of the shipyard and the dry docking facility was updated on 9/11/2001
- Float barriers required around naval ships in yard
- Float barriers or barges required around dry docks or ends of dry docks with ships in them.
- Afloat patrols

References

- Mil-Std-1625D, Certification of Dry Docking Facilities, USN
- ABS Rules for Building and Classing Steel Floating Dry Docks 2009; Publication 11 Dry Docks
- SFLC Standard Specification 8634
- Docking Instructions and Routine Work in Dry Dock: NSTM 997
- Safe Operation and Maintenance of Dry Dock Facilities; ASCE Engineering Practice No. 121
- Guide to Marine Forensic Investigations; SNAME T&R 8-1
- Marine Casualty Response
- Dry Dock Training Manuals (numerous)

AUTHOR INDEX